Catalogue No. 23
1925

The Bridgeport Hardware Manufacturing Corporation

Bridgeport, Connecticut

U. S. A.

Manufacturers of

High Grade Hand Tools and Hardware Specialties

Cable Address, "SUREGRIP," Bridgeport

Codes: Western Union Universal, Commercial Telegraph,
Bentley's Complete Phrase

Sales Representatives

UNITED STATES

J. C. McCarty & Co..29 Murray St., New York City
C. W. Gause Co..693 Mission St., San Francisco, Cal.
Dan. M. Bell..Slaughter Bldg., Dallas, Texas
F. M. Furlong..190 N. State St., Chicago, Ill.

CANADA

Sutherland Bros..Confederation Life Bldg., Winnipeg
H. A. Harrison..220 King St., West, Toronto (Province of Ontario)

Foreign Sales Representatives

AFRICA—Union of South Africa & Rhodesia.
    Cape Town—Brittain & Emery, P. O. Box 1665.
    Johannesburg—Brittain & Emery, P. O. Box 6398.
ARGENTINA—M. J. Garcia & Cia., Calle Lima 486, Buenos Aires, Argentina.
ASIA—Muller & Phipps (Asia) Ltd., 25 West 44th St., New York, N. Y. (Home Office.)
AUSTRALIA—Max Steinberg, Macdonnell House, Sydney, Australia.
CEYLON—Muller & Phipps (Asia) Ltd., 46 Keyser St., P. O. Box 304, Colombo, Ceylon.
CHINA—Hongkong—Muller & Phipps (Asia) Ltd., Princes Bldg., P. O. Box 25.
    Shanghai—Muller & Phipps (China) Ltd., No. 2 Canton Road, P. O. Box 650.
EGYPT—Muller & Phipps (Asia) Ltd., 71 Chareh Sekkah-el-Guediah, P. O. Box 1764, Cairo, Egypt.
FINLAND—C. K. Hornsleth & Nissen, Brunkebergstorg 24, Stockholm, Sweden.
HAWAII—Muller & Phipps (Hawaii) Ltd., cor Fort & King Sts., Honolulu, Hawaii.
INDIA—Bombay—Muller & Phipps (India) Ltd., 14-16 Greene St.
    Calcutta—Muller & Phipps (India) Ltd., 21 Old Court House St., P. O. Box 263.
    Karachi—Muller & Phipps (India) Ltd., 1024 Napier Road.
    Madras—Muller & Phipps (India) Ltd., 21 Sunkurama Chetty St.
    Rangoon, Burma—Muller & Phipps (India) Ltd., 4-5 Shafraz Rd., P. O. Box 604.
JAMAICA—Ben Alberga & Co., 70 King St., P. O. Box 50, Kingston, Jamaica.
JAPAN—Osaka—Muller, Phipps & Sellers, Ltd., Dojima Building, P. O. Box 63.
    Tokyo—Muller, Phipps & Sellers, Ltd., Marunouchi Building, Central P. O. Box 98.
MEXICO—Padilla Bros. & Bethencourt, 23 Uruguay Ave., Apartado 1676, Mexico, D. F.
NEW ZEALAND—Max Steinberg, Wellington, P. O. Box 893.
PHILIPPINE ISLANDS—Muller & Phipps (Manila) Ltd., Pacific Bldg., P. O. Box 349, Manila, Philippine Islands.
PORTO RICO—Felix A. Aguilar, P. O. Box 1222, Tetuan No. 37, San Juan, P. R.
STRAITS SETTLEMENTS—Muller & Phipps (Malaya) Ltd., 4 Cecil St., Singapore, Straits Settlements.
SWEDEN—C. K. Hornsleth & Nissen, Brunkebergstorg 24, Stockholm, Sweden.
SWITZERLAND—Harris Cohen, Lutry (Lausanne), Switzerland.
URUGUAY—M. J. Garcia & Cia., Buenos Aires, Argentina.

OUR PLANT AT IRASTAN AVE., ADMIRAL AND WASHBURN STREETS, BRIDGEPORT, CONN.

Terms

---

30 days net.   Two per cent. discount for Cash in 10 days.
F. O. B. cars Bridgeport.

Accounts not paid when due will be subject to sight draft.

Net prices will be quoted upon application.

All prices are subject to change without notice.

The Unbreakable Nail Puller

No. 20

Every One Tested and Warranted

Unbreakable Rammer. The rammer of this puller is made of **Malleable Iron** which cannot be broken under the hardest strain to which a nail puller is subjected.

Oval Shape. This Nail Puller will not roll as the Rammer is oval and the shank, being rectangular in shape, cannot turn in the handle.

Detachable Claw. The Nail Puller Claw bears the brunt of the work and on the **"Unbreakable"** Puller, the claws are interchangeable and can be easily detached and replaced when worn out.

Box Joint. This construction keeps the jaws always in alignment so they grip the nail centrally and do not slip off. **It also gives great strength and long wear.**

Hand Guard. It is impossible to pinch the hand with this puller. The improved hand guard on the shank gives absolute protection.

The "Unbreakable" Puller is very attractively finished. All steel parts are polished. The Rammer is enameled jet black and the handle or grip finished in red enamel.

All materials used in the **"Unbreakable"** Nail Puller are the finest obtainable for the purpose. The jaws are very carefully hardened and tempered by skilled workmen and every puller is carefully tested before it leaves the factory.

Size, 18 inch.   Weight, 4¾ lbs.   **List Price, $30.00 per dozen**

Packed 3 dozen in a case.

Size of case, 2½ cubic feet.

Gross weight, 192 lbs.   Net weight, 169 lbs.

Tiger Nail Puller

No. 48

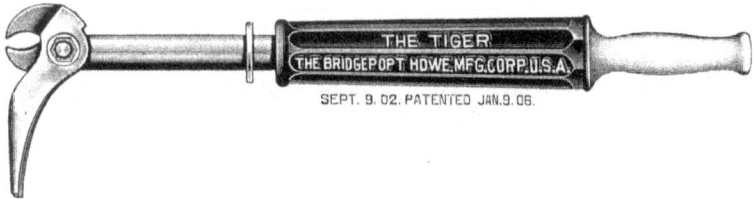

Every Puller Tested and Warranted

Shank. High grade special steel demonstrated by years of testing to be the best quality for the purpose.

Claw. Made from same quality steel as shank, forged from bar steel in specially constructed dies which insure absolute uniformity of size for a perfect fit with the forged end of shank.

All wearing parts of both shank and claw are milled to gauge and hardened to a uniform temper. The claw is so constructed that when assembled with shank, we have a box joint, giving a double shoulder bearing of claw on shank, and insuring true alignment of jaws.

Hand Guard. Securely fastened on shank affords protection to hands of users.

Claw is fastened to shank with bolt and nut, and is interchangeable.

Handle or Rammer is reinforced at all points subject to strains, and cored perfectly smooth inside so that the shank will move freely in any direction.

Finish. Handle end of Rammer is aluminum finish, the body a glossy black enamel, baked on.

Size, 18 inch.   **List Price, $24.00 per dozen**

Packed 3 dozen in a case.

Size of case, $2\frac{1}{4}$ cubic feet.

Gross weight, 192 lbs.   Net weight, 174 lbs.

New Jumbo Nail Puller
No. 57

PATENTED JAN. 9.1906.

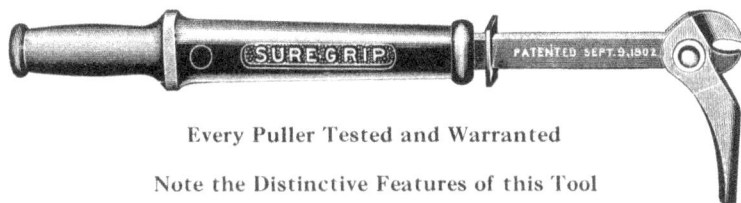

Every Puller Tested and Warranted

Quality and Workmanship. A high grade tool, jaws and shank forged from the finest quality tool steel. Forging, hardening and tempering done by men of long experience.

Construction and Finish. Hand guard protects the user's hand. Claw fastened to shank with a bolt and nut which permits tightening when loose from wear. New style frictionless Rammer so cored that it works smoothly. Steel parts polished, Rammer finished in jet black enamel.

Size, 18 inch.   List Price, $18.00 per dozen

    Packed 3 dozen in a case.   Size of case, 1¾ cubic feet.

Gross weight, 194 lbs.   Net weight, 175 lbs.

Suregrip Nail Puller
No. 56

Every Puller Tested and Warranted

Note the Distinctive Features of this Tool

Square Shank with improved hand guard. Claw is automatic having no spring.

Double Shoulder Bearing of claw on shank gives great strength and long wear.

Box Joint. This construction keeps claws always in alignment. Nails are gripped centrally.

Broad Foot. That portion of the claw which rests on box lids, has a broad bearing surface which prevents splitting thin box lids.

Handle finished in jet black enamel, other parts natural steel finish with points of jaw polished bright.

Size, 18 inch.   Weight, 4¼ lbs.   List Price, $16.50 per dozen

    Packed 3 dozen in a case.   Size of case, 2¼ cubic feet.

Gross weight, 168 lbs.   Net weight, 150 lbs.

Red Bull Nail Puller
No. 54

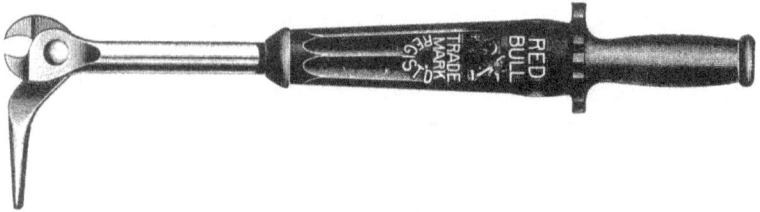

Every Puller Tested and Guaranteed

Shank and Claw are forged from a fine grade of special steel, hardened and tempered by expert workmen. Foot of claw is flat and broad, which prevents splitting of thin box lids.

Rammer shaped to give the maximum of strength. The Red Bull will not roll on account of the four knobs just below the handle, which also provide an easy rest for the hand.

Finish. Very attractively finished, the rammer being jet black with the grip enameled in bright red. Shank and claw are nicely polished.

Size, 18 inch.   **List Price, $18.00 per dozen**

Packed 3 dozen in a case.   Size of case, 1¾ cubic feet.

Gross weight, 200 lbs.   Net weight, 182 lbs.

Rex Nail Puller
No. 64
Trade Mark Registered in Japan, May 1, 1924
Certificate No. 160453

A Reliable Tool

Jaws and shank forged from good quality steel, hardened and tempered.

Jaws and shanks, black oil finish, with points fine polished, handles finished in jet black enamel.

Size, 18 inch.   **List Price, $15.00 per dozen**

Weight, 4¼ lbs.

Packed 3 dozen in a case.   Size of case, 1¾ cubic feet.

Gross weight, 172 lbs.   Net weight, 153 lbs.

The Tomahawk Tool

A HATCHET—HAMMER—NAIL PULLER—BOX CHISEL

A Handsome Three-Color Lithographed Steel Display Stand as shown below, packed with each ½ dozen Tomahawks

A "Silent Salesman" That Will Earn Good Profits

One of the Handiest Tools ever made for use in Store, Office, Warehouse or Home

The **TOMAHAWK** is really four tools in one; all of them good, and its many uses are too numerous to mention. It is built just right to stand use and abuse wherever there is CUTTING, PRYING, NAIL PULLING or HAMMERING to be done.

Made all in one piece from high quality carbon steel, drop-forged, properly hardened and tempered. Fitted with comfortable oval hardwood scale handle, securely riveted. This solid construction makes THE TOMAHAWK practically INDESTRUCTIBLE.

Each tool packed in individual carton

No. 99-S   TOMAHAWK—Packed ½ dozen in box with Display Stand.
No. 99   TOMAHAWK—Without Display Stand.

List Price, $18.00 per dozen

Length, 13 inches.	Weight, 24 ounces.	Display Stand, 12 ounces.

		Gross weight	Net weight	Volume of case
Packed 6 dozen in a case.				
No. 99-S	(with stands)	172 lbs.	110 lbs.	6½ cubic feet
No. 99	(without stands)	140 lbs.	110 lbs.	4½ cubic feet

FULLY GUARANTEED

Tomahawk Jr. Tool
No. 9

"The Handy Home Tool"

Drop forged from a high grade steel and is practically indestructible.

Useful as a nail puller—box opener—ice chopper—tack puller—hammer—nut cracker and many other odd jobs.

Gun metal finish—Cutting edge, claws and hammer face polished.

Length, 9¼ inches.   Weight, 9 ounces each.

Individual boxes.   12 dozen in a case.

List Price, $6.00 per dozen

Gross weight of case, 107 lbs.   Volume, 2.2 cubic feet.

Matchless Hatchet
No. 85

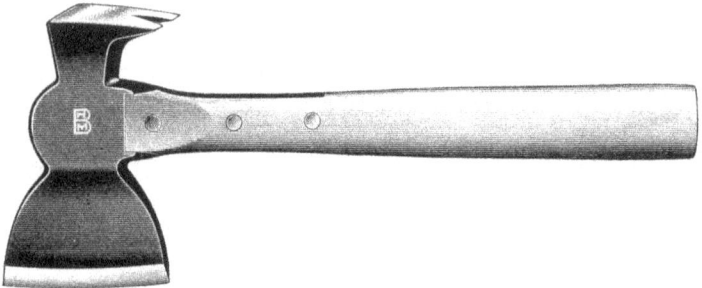

"The Hatchet That Never Loses Its Head"

A general utility hatchet of unique construction embodying the first real improvement in years.

Head and shank forged in one piece and held securely to hickory handle by means of rivets and cannot possibly come off or loosen—a common fault with all other kinds. Note the Clip Claw for pulling nails.

Drop forged from fine quality steel and carefully hardened and tempered and WILL HOLD ITS EDGE.

Gun metal finish—Cutting edge and face of hammer polished.

Weight, 26 ounces each.   Length over all, 13 inches.

List Price, $24.00 per dozen

6 dozen in a case.   Volume, 5 cubic feet.

Gross weight, 156 lbs.   Net weight, 126 lbs.

The Matchless Opener

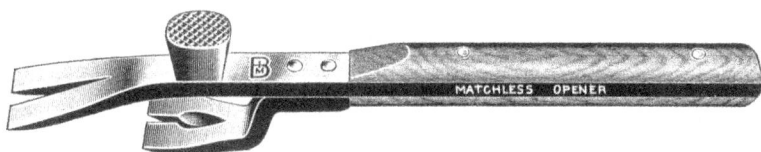

Patented Sept. 29, 1908

A Tool of Great Usefulness and Strength Combining Chisel, Hammer and Nail Puller

An exceptionally good tool for opening boxes, crates, barrels, etc., and very handy for many other purposes.

Made of high grade drop forged steel. The hardwood scales, securely riveted on, form an oval handle which fits the hand perfectly. It will stand the hardest use and abuse for years.

Every one guaranteed

The cupped cleft in the lower claw is a feature. Nails cannot slip out from the cleft when pulling them.

Length, 13½ inches.　　　Weight, 26 ounces each.

No. 90.　Natural steel finish, polished claws.

List Price, $12.00 per dozen

No. 92.　Full polished all over.

List Price, $15.00 per dozen

Weight, 20 lbs. per dozen net.

Packed ⅓ dozen in a box.　　　6 dozen in a case.

Size of case, 2 cubic feet.

Gross weight, 141 lbs.　　　Net weight, 120 lbs.

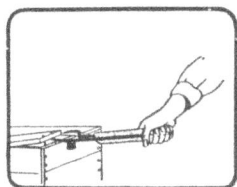

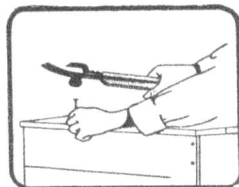

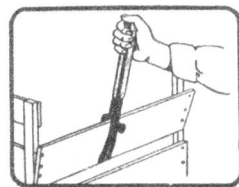

The Seminole Opener

No. 120

Combining a Hammer, Box Chisel, Hatchet, and a Cutter for Opening Corrugated Paper Cartons and Cases

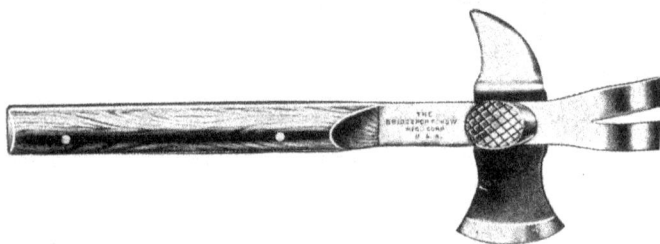

Every one warranted

The Seminole is made of drop forged steel with hardwood scales securely riveted to the main portion, forming an oval handle which fits the hand perfectly and will never loosen.

The hatchet blade is made of fine, high carbon steel, tempered just right to hold its edge.

The solid construction and the use of fine material makes the Seminole Opener practically unbreakable.   It will stand the hardest abuse.

Length, 13 inches.   Weight, 22 ounces.   Packed 6 dozen in a case.

List Price, $18.00 per dozen

Size of case, $6\frac{1}{2}$ cubic feet.   Gross weight, 163 lbs.   Net weight, 138 lbs.

Forged Steel Box Chisel

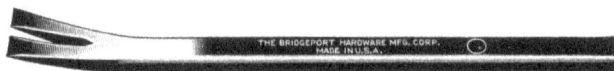

Forged from Bar Steel, Hardened and Tempered

No. 10.   Polished all over.   **List Price, $6.00 per dozen**

No. 20.   Black Steel finish, Points Polished.   **List Price,   5.00 per dozen**

Size, $\frac{1}{2}$ x $\frac{3}{4}$ x $13\frac{1}{2}$ inches.

Packed $\frac{1}{2}$ dozen in a box.   10 dozen in a case.

Size of case, $1\frac{1}{2}$ cubic feet.

Gross weight, 163 lbs.   Net weight, 146 lbs.

The Karton Kutter

No. 14

A new tool of great usefulness needed many times a day in every store

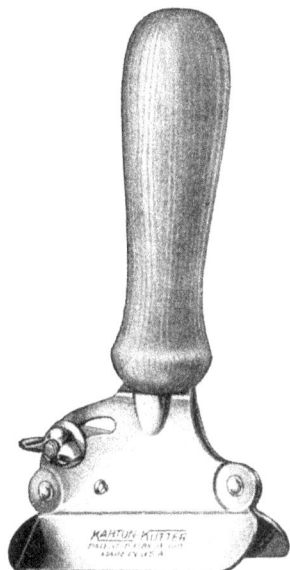

BLADE CAN BE
TAKEN OUT TO
RESHARPEN

ADJUSTABLE FOR
CUTTING THICK
OR THIN CARTONS

BLADE LOCKS IN
ANY POSITION

**BLADE FOLLOWS CORNERS OR
CUTS STRAIGHT ACROSS BOX AS
DESIRED BUT DEPTH OF CUT IS
REGULATED — INSIDE BOXES
ARE PROTECTED.**

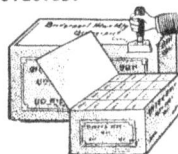

The first really practical device for opening fibre or corrugated paper shipping containers

It opens all kinds quickly and neatly. Three motions as shown and they open "like a book."

Leaves cartons in fine condition to use again.

Cutting blades made of razor blade steel, perfectly hardened and tempered and will give long service, and can be resharpened.

Perfectly safe to handle—sharp edge is protected on all sides.

Handle and frame securely riveted together. Extra strong construction throughout.

Very attractively finished—full nickel plated and buffed frame, red varnished hardwood handle.

Packed individually.

6 dozen to a case.	Length over all, 6 inches.	Weight, 6 ounces each.
Gross weight, 46 lbs.	Net weight, 20 lbs.	Volume, 2½ cubic feet.

List Price, $1.00 each

Perfection Scraper
No. 16

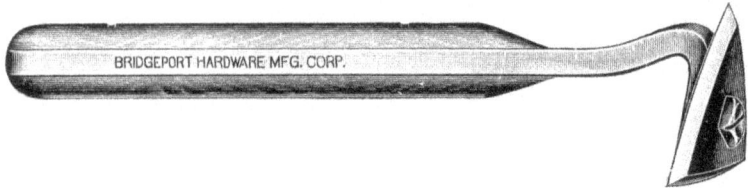

The construction of this tool is ideal for a box scraper. The handle is formed by riveting two hardwood scales to the steel bar to which the blade is fastened.

This construction provides an oval handle which is very strong and will never loosen. It gives weight to the tool, which is a big advantage.

A heavy box scraper "takes hold" and cuts easily, requiring little pressure or power from the operator.

Blade is made of high grade tool steel, carefully hardened and tempered.

It can be quickly turned to bring either of the three cutting edges into play, or removed for sharpening.

Width of blade, 3 inches.

Length over all, 13 inches; weight, 24 ounces.   **List Price, $12.00 dozen**

Extra blades for above.   **List Price,   3.00 dozen**

Packed 6 dozen in a case.
Size of case, 2½ cubic feet.

Gross weight, 128 lbs.   Net weight, 108 lbs.

The Dandy Box Scraper
No. 15

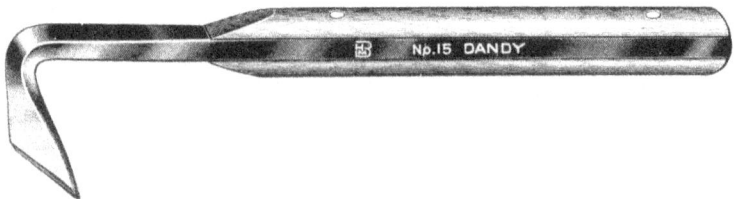

Similar in construction to the Perfection Scraper shown above but with only one cutting edge.

Width of blade, 2 inches.

Length, 12½ inches.   Weight, 20 ounces each.

List Price, $9.00 per dozen

Packed ½ dozen in box.   6 dozen in a case.

Size of case 1½ cubic feet.

Gross weight, 114 lbs.   Net weight, 96 lbs.

The 20th Century Hammer
No. 105
A Combination Hammer and Opener for Boxes and Barrels

The 20th Century Hammer takes the place of an ordinary nail hammer and is much more useful for odd jobs. It will outlast two ordinary hammers.

The handle won't break—solid steel from end to end—the head won't loosen.

The offset claw pulls nails straight.

It is made of forged steel with both claws and hammer head hardened and tempered. Two hardwood scales riveted to the steel portion form an oval handle. This construction makes the tool strong enough to withstand the hardest use and abuse.

Length, 13 inches.   Weight, 24 ounces.

Packed 6 dozen in a case.   Size of case, 2¾ cubic feet.

List Price, $12.00 per dozen

Gross weight, 134 lbs.   Net weight, 114 lbs.

The Iroquois Hatchet Tool
No. 115
Combining a Hatchet, Hammer and Box Opener

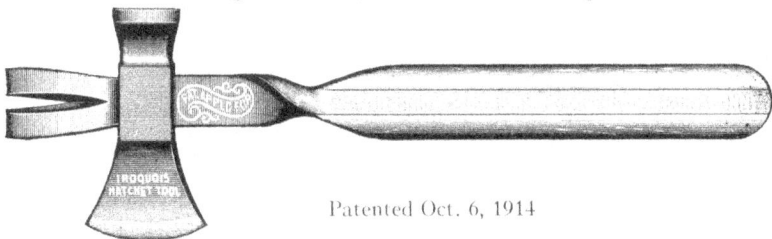

Patented Oct. 6, 1914

The hatchet blade is made of high carbon steel, carefully tempered and forged around a solid bar of steel, to which are riveted hardwood pieces forming an oval handle. Neither the head (hatchet and hammer) nor the handle can come off or work loose on account of the patented construction. The steel portions are oil finished, with hatchet edge, claw and face of hammer full polished. Wood portions of handle are natural wood color and varnished.

Length, 12½ inches.   Weight, 24 ounces.   Packed 6 dozen in a case.

List Price, $15.00 per dozen

Size of case, 2½ cubic feet.   Gross weight, 160 lbs.   Net weight, 138 lbs.

The Hooker Crate Opener

The best pocket tool for opening boxes, crates, barrels, prying off lids, hammering, etc.

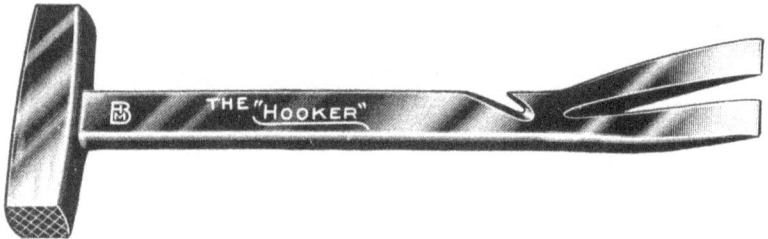

Patented April 21, 1914

Every one warranted

An exclusive patented feature of this tool is the cleft or "hook" in the side. This makes it effective for pulling nails which cannot be drawn with ordinary crate openers.

A nail can be gripped in most any position and at just the right angle to pull easily.

One face of hammer-head scored, one plain polished.

Weight, 15 ounces.   Length, 9 inches.

An Exclusive New Feature and a Big Improvement

The Hooker Opener is forged from a bar of fine quality steel, all in one piece. It is very carefully hardened and tempered and finely finished.

No. 40.   Natural finish, with claw and face of head polished.   **$9.00 per doz.**

No. 42.   Full polished, heavily nickel plated and buffed.   **12.00 per doz.**

Packed ½ dozen in a box.   6 dozen in a case.

Size of case, 1¼ cubic feet.

Gross weight, 85 lbs.   Net weight, 68 lbs.

Box Terrier Crate Opener

Warranted

This Tool is Forged from a Bar of Fine Quality Steel all in One Piece

The Box Terrier was designed after careful study and the shape is exactly right. Easy on the hand and pocket. All edges are rounded.

Claw end hardened, tempered and tested. Hammer head hardened and tempered. One face scored, one smooth. Packed ½ dozen in box.

No. 80. Natural steel finish, polished claw. Weight, 15 oz. Length, 9¼ inches.

List Price, $9.00 per dozen

No. 82. Nickel plated and buffed. Weight, 15 oz. Length, 9¼ inches.

List Price, $12.00 per dozen

Packed 6 dozen in a case.   Size of case, 1 cubic foot.
Gross weight, 83 lbs.   Net weight, 68 lbs.

No. 96. Natural steel finish, polished claw. Weight, 10 oz. Length 8¼ inches.

List Price, $8.00 per dozen

No. 98. Nickel plated and buffed. Weight, 10 oz. Length, 8¼ inches.

List Price, $10.00 per dozen

Packed 6 dozen in a case.   Size of case, 1 cubic foot.
Gross weight, 62 lbs.   Net weight, 50 lbs.

Nox Tox Crate Opener

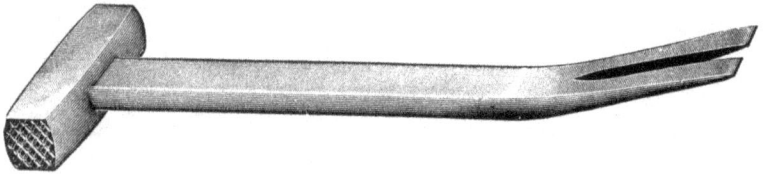

Forged in One Piece.   Warranted

Length, 9¼ inches over all.   Weight, 15 ounces each.

Hardened head, tempered claw.   One hammer head scored, one plain.

No. 70. Natural steel finish, polished claw.   **List Price, $6.00 per dozen**

No. 72. Full polished all over.   **List Price, 10.00 per dozen**

Packed ½ dozen in box.   6 dozen in case.

Size of case, 1 cubic foot.

Gross weight, 83 lbs.   Net weight, 68 lbs.

Baby Terrier Crate Opener

A Perfect Pocket Tool

Just the right weight for light work and very strong. Forged from the bar in one piece, carefully hardened and tempered. Fits the pocket perfectly—no sharp edges.

Warranted

Length, 6½ inches.   Weight, 8 ounces.

No. 76. Natural steel finish, polished claw.   **List Price, $6.00 per dozen**

No. 78. Highly nickel plated and buffed.   **List Price,   9.00 per dozen**

Packed 6 dozen in a case.

Size of case, ¾ cubic foot.

Gross weight, 48 lbs.   Net weight, 36 lbs.

The Giraffe Reacher
No. 138

A perfect device for taking down bottles, cans, lamp chimneys, etc., from high shelves in a store and bringing them safely to the counter. It is also very convenient for dressing show windows.

The **Giraffe** can be held and operated with one hand. The jaws open automatically and are closed by pressing the hand lever—a simple locking device in the middle of the handle locks the jaws at any point.

The steel jaws of the Giraffe are covered with heavy rubber which eliminates slipping when handling glassware, etc.

A wonderfully effective device—like a human hand on an arm four feet long.

Individual Packing. Each reacher in a heavy corrugated carton suitable for shipping singly. **List Price, $18.00 dozen**

50 in a case. Net weight, 143 lbs. Gross weight, 188 lbs. Volume, 19 cubic feet.

Bulk Packing. Especially for export, 100 reachers nested in a case.

Gross weight, 320 lbs. Net weight, 275 lbs. Volume, 15 cubic feet.

Matchless Screw Drivers

The blades of Matchless Screw Drivers have enormous strength and will stand the most severe prying and twisting strains.

Handles are large in diameter—you can get a greater grip with a Matchless Screw Driver and turn any screw within blade capacity without cramping or tiring either wrist or hand.

Driving 8 Inch Through Hardwood Timber

8 Inch Size Supporting Weight of Two Men

The steel used in Matchless drivers is of a special analysis, found after many tests and years of experience to be the best for this purpose. Selected wood of great toughness and strength is used for the handles.

Heat Treatment

All blades are heat treated by a special process making the points extremely tough and the shanks "springy" so they can be subjected to very severe strains without bending, when used for prying, etc. (See cuts above.)

Finish

All steel parts finely polished. Handles have a smooth dull black finish which will not crack or chip.

Reliability

Matchless Screw Driver blades are tested twice before assembly under greater pressure than the hand can apply. Unconditionally guaranteed to be free from defects in workmanship or material.

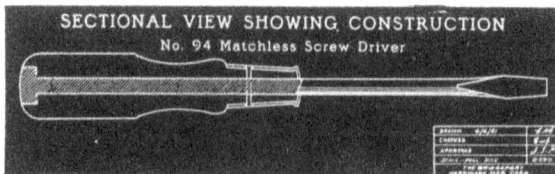

SECTIONAL VIEW SHOWING CONSTRUCTION
No. 94 Matchless Screw Driver

Matchless Screw Driver
No. 92

Blade securely held by special winging and by a steel pin through blade, ferrule and handle.

Size, inches,		2	2½	3	4	5	6	8	10	12
List Price, per dozen,		$3.00	3.25	3.50	4.25	5.00	6.00	8.00	10.00	12.00

Matchless Machinists' Screw Driver
No. 93

Made with square blade so it can be turned with wrench or pliers when great power is needed.

Steel extends clear through the handle and is held securely by a steel rivet through ferrule, blade and handle and by the hexagonal shape at butt.

Steel in butt is extra thick. They can be hammered on without injury to the driver.

Size, inches,		3	4	5	6	8	10	12
List Price, per dozen,		$3.50	4.25	5.00	6.00	8.00	10.00	12.00

Matchless Mechanics' Screw Driver
No. 94

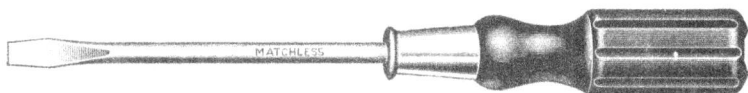

The blade extends through the handle and is held securely by the hexagonal shape at butt and by pin through ferrule, blade and handle.

The thickness of steel in butt permits hammering on it without injury to the driver.

Packed ½ dozen in a box or one dozen assorted with display stand as shown on page 28.

Size, inches,		2	2½	3	4	5	6	8	10	12
List Price, per dozen,		$3.00	3.25	3.50	4.25	5.00	6.00	8.00	10.00	12.00

All styles and sizes of Matchless Screw Drivers packed ½ dozen in box.

Matchless Electricians' Screw Driver
No. 95
Insulated Handle

Blade runs clear through handle and is covered at the end by an insulator protected by a steel cap which also protects the handle.

Size, inches,	2	2½	3	4	5	6	8	10	12
List Price, per dozen,	**$3.00**	**3.25**	**3.50**	**4.25**	**5.00**	**6.00**	**8.00**	**10.00**	**12.00**

Matchless Cabinet Screw Driver
No. 96

Blade ³⁄₁₆ inch in diameter and extends through the handle and is held securely by pin and hexagonal shape at butt. Point is shaped correctly for reaching recessed screws.

Size, inches,	2½	3½	4½	5½	6½	8½	10½	12½
List Price, per dozen,	**$3.00**	**3.50**	**4.25**	**5.00**	**6.00**	**8.00**	**10.00**	**12.00**

Matchless Heavy Machinists' Screw Driver
No. 97

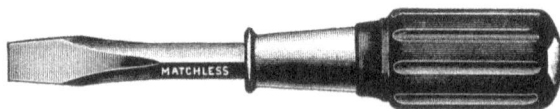

Heavy square blade. Wrench or pliers can be applied for greater power. Steel extends through handle and thick steel at butt can be hammered on without injury to the driver.

Length of Blade, inches,	2½	3½	4½	5½
Diameter, inches,	³⁄₈	³⁄₈	⁷⁄₁₆	⁷⁄₁₆
List Price, per dozen,	**$5.00**	**6.00**	**7.00**	**8.00**

All styles and sizes of Matchless Screw Drivers packed ½ dozen in box.

Matchless Electricians' Screw Driver
No. 98

Designed Especially for Electricians' Use

Blade $\frac{3}{16}$ inch in diameter (all lengths). Point is the same width as the shank, so recessed screws can be reached and turned. Magnetized points. Blade, ferrule and handle securely riveted together.

Size, inches,	$2\frac{1}{2}$	$3\frac{1}{2}$	$4\frac{1}{2}$	$5\frac{1}{2}$	$6\frac{1}{2}$	$8\frac{1}{2}$	$10\frac{1}{2}$	$12\frac{1}{2}$
List Price, per dozen,	$3.00	3.50	4.25	5.00	6.00	8.00	10.00	12.00

Matchless Baby Screw Driver
No. 99

A Fine Little Tool for Small Work. Blade Extends Clear Through Handle.

Length over all, 4 inches. Blade, $\frac{3}{16}$ x $1\frac{1}{2}$ inches.

List Price, $3.60 per dozen

Matchless "Little Bully" Screw Driver
No. 970

For Heavy Work at Close Quarters

A powerful tool which can be used in places where space is limited and the ordinary screw driver is useless. Measures only 6 inches over all. Blade extends clear through the handle. It can be used to hammer **on** and hammer **with,** and will stand the roughest use and abuse. Made in one size only.

Length over all, 6 inches. Blade, $\frac{3}{8}$ x $1\frac{3}{4}$ inches.

List Price, $6.00 per dozen

All Matchless Screw Drivers packed $\frac{1}{2}$ dozen in a box.

Matchless Shock Proof Screw Driver
No. 100
Designed especially for electrical work

Mechanics in every line find it an exceptionally good screw driver for all kinds of light work and superior to the ordinary wood handle type.

Handles of special composition rubber moulded solidly around the shank of the blade eliminating the possibility of the handle becoming loose or turning. Tough and durable and will not crack or chip.

Blades of a special high grade steel finely polished and tempered the entire length. Point is the same width as shank so recessed screws may be reached and turned. Tested to withstand 18,000 volts of electricity. Blades magnetized—a feature appreciated by the user.

Packed ½ dozen in box.

Size, inches,	2½	3½	4½	5½	6½	8½	10½	12½
List Price, per dozen,	**$5.15**	5.45	5.80	6.30	6.90	7.80	9.30	11.00

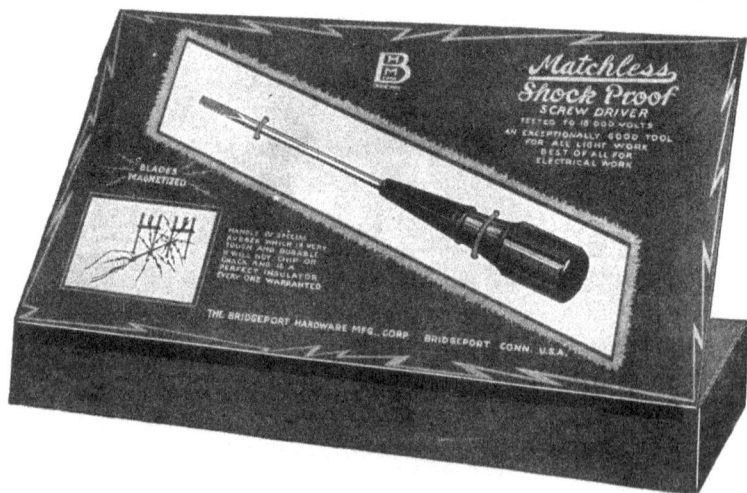

No. 60 Assortment

Comprises the following sizes:

2—3½ inch.   3—4½ inch.   2—5½ inch.   3—6½ inch.   2—8½ inch.

Packed in an attractive display box finished in three colors—Black, Red and Gold.   A SILENT SALESMAN of real ability.

Each Assortment in an individual container box.

Weight, 2¾ lbs. each.   18 in case.   Size of case, 3½ cubic feet.

List Price, $7.20 each

Challenge Mechanics' Screw Driver
No. 32

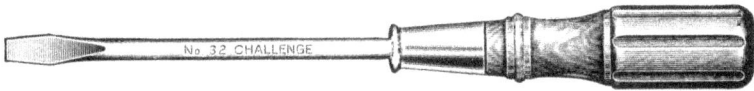

Fully Warranted

Round blade, forged from special analysis steel and tempered from ferrule to point. Heavy polished steel ferrule. Ferrule, blade and handle securely fastened together by steel pin. Fluted handle, stained dark red and varnished.

Every blade tested twice before assembly.

Size, inches,	2	2½	3	4	5	6	8	10	12
List Price, per dozen,	$2.00	2.25	2.50	3.00	3.50	4.00	5.50	7.00	9.00

Packed ½ dozen in box.

Challenge Electricians' or Cabinet
No. 36

For Electricians' or Cabinet Makers' Use

Extra fine steel blade, tempered from ferrule to point. Polished steel ferrule. Red fluted varnished handle. Blade, ferrule and handle fastened together by flush rivet. Blades are magnetized—a feature appreciated by mechanics. Diameter in all sizes, $\frac{3}{16}$ inch.

Size, inches,	2½	3½	4½	5½	6½	8½	10½	12½
List Price, per dozen,	$2.25	2.50	3.00	3.50	4.00	5.50	7.00	9.00

Packed ½ dozen in box.

Baby Challenge Screw Driver
No. 37

Polished blade and ferrule. Fluted handle, stained cherry red and varnished.

Length over all, 4 inches.   Blade, $\frac{3}{16}$ x 1½ inches.

List Price, $3.00 per dozen

Packed ½ dozen in box.

Giant Challenge Screw Driver
No. 38

Made especially for Millwrights, Engineers, etc., for long reaches and extra heavy work. Extra heavy, bull neck steel ferrule. Round blade ½ inch in diameter and fastened with steel pin through handle and ferrule.

Fluted handle with double grip. Red varnished finish. All steel parts highly polished. Every one warranted.

Size, inches,	15	18	24
List Price, per dozen,	**$15.00**	**18.00**	**24.00**

Packed ½ dozen in box.

Vulcan Mechanics' Screw Driver
No. 252

Octagonal Tool Steel Blade

Blade forged from special octagonal steel, hardened and tempered from point to ferrule, natural black finish, with flat sides of point highly polished. Hardwood handle, black "Rubberoid" finish, nickel plated ferrule, polished and buffed. Blade, ferrule and handle are securely riveted together with heavy steel pin.

The blade in this driver is exceptionally strong and springy. It will withstand an enormous prying strain, and spring back straight and true if bent.

Size, inches,	2	2½	3	4	5	6	8	10	12
List Price, per dozen,	**$3.00**	**3.25**	**3.50**	**4.25**	**5.00**	**6.00**	**8.00**	**10.00**	**12.00**

Packed ½ dozen in box.

Vulcan Cabinet Screw Driver
No. 254

Octagonal Tool Steel Blade

Same description as Vulcan Mechanics' Screw Driver shown above but with smaller blade for Electricians or Cabinet Makers. Blade 3_{16} inch in diameter on all sizes.

Size, inches,	2½	3½	4½	5½	6½	8½	10½	12½
List Price, per dozen,	**$3.25**	**3.50**	**4.25**	**5.00**	**6.00**	**8.00**	**10.00**	**12.00**

Packed ½ dozen in box.

Hercules Knife Handle Screw Driver
No. 102

Patented August 26, 1913

Superior for Motorists, Repairmen, Plumbers, etc., who use screw drivers for prying, etc.

An improved tool in the popular knife handle type. The blade is made of high grade special steel, heat treated by a method that makes it wonderfully strong and springy. It is fastened into the handle by a patented process giving the same strength in the handle as the one piece drivers of this pattern, but **far greater strength in the blade, which is the all important part.**

For those who use screw drivers for extra hard jobs—prying with the blade, hammering on the butt, chiseling, etc.—the Hercules is in a class by itself. It will "stand up" under more hard use and abuse than any other screw driver.

Wood scale handle, natural wood polished.   Waterproof finish.

Size, inches,	2	3	4	5	6	8	10	12
List Price, per dozen,	$3.50	4.00	4.25	5.00	6.00	8.00	10.00	12.00

Packed ½ dozen in box.

Hercules Knife Handle Screw Driver
No. 12

Heavy Machinists' Pattern

Especially Designed for Heavy Work

Square shank permits turning it with a wrench when great power is needed.

Length of blade, 4½ inches.   Length over all, 9½ inches.   Size of square blade, 7⁄16 inch.   Point, 9⁄16 inch wide.

List Price (one size only, 4½ inches), **$12.00 per dozen**

Weight per dozen, 9 lbs.

Matchless Screw Driver

No. 94

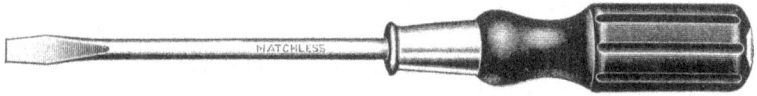

Blades are made of special analysis steel and are tempered from ferrule to point by a special process. They will withstand an enormous prying and twisting strain.

Steel extends clear through and the blade cannot turn in the handle. It is held securely by a heavy steel rivet through ferrule, blade and handle and by the hexagonal shape at butt.

Handles large in diameter giving greater power for turning screws and will not cramp and tire the hand. Packed ½ dozen in box.

Size, inches,	2	2½	3	4	5	6	8	10	12
List Price, per dozen,	$3.00	3.25	3.50	4.25	5.00	6.00	8.00	10.00	12.00

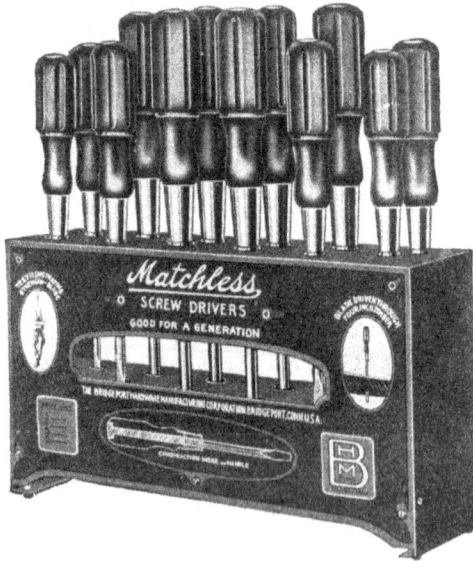

No. 40 Assortment

Comprises the following sizes: 2—3 inch. 4—4 inch. 2—5 inch. 4—6 inch.

The stand is made entirely of welded steel enameled in four colors. It displays the Matchless Screw Drivers in a very attractive manner and is a REAL SILENT SALESMAN showing the points of merit clearly and quickly. This display stand alone is worth a dollar to any dealer. It pleases critical merchants who will give no space to anything but the finest dealer helps.

Note the rubber feet. It won't scratch. Each assortment in an individual container box.

Weight, 6 lbs each.   24 in case.   Size of case, 6½ cubic feet.

List Price, $6.00 each

Reliance Screw Driver
No. 7

An Exceptionally Fine Driver at a Moderate Price

Every One Warranted

Blade—Made of special steel. If bent under pressure it will spring back straight and true.
Handle—Made of hardwood, fluted design.
Ferrule—Made of heavy drawn steel.

Construction—Handle, blade and ferrule securely riveted together.
Finish—Fine, polished blade, nickel plated ferrule, highly buffed, black rubberoid handle.

Size, inches,	2	2½	3	4	5	6	8	10	12
List Price, per dozen,	$2.95	3.15	3.32	3.56	4.00	4.25	6.65	7.80	9.32

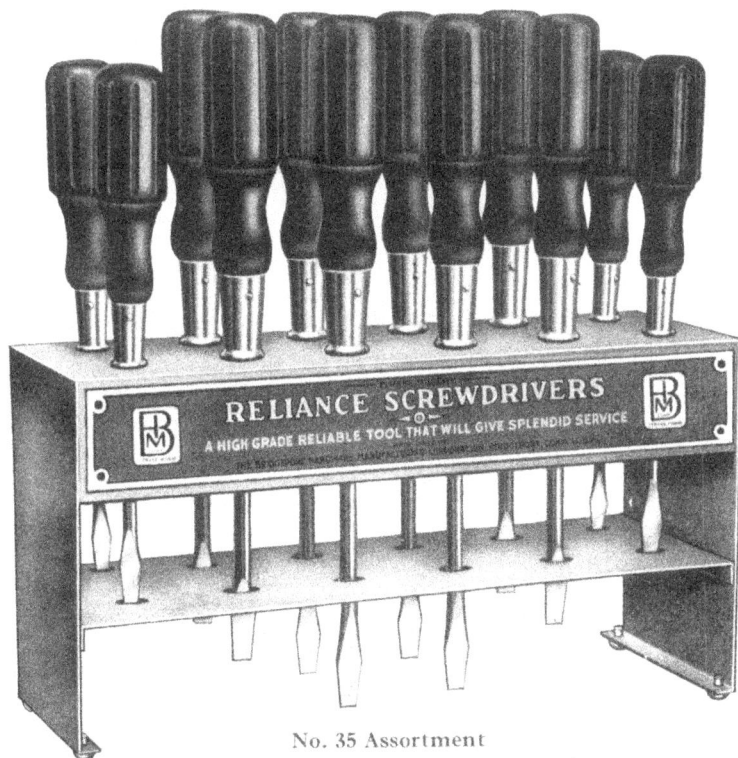

No. 35 Assortment

This fine driver is packed when desired in a very attractive display rack containing the following drivers: four 4 inch, four 5 inch, and four 6 inch.

The stand is made entirely of steel electrically welded and is very attractively enameled in four colors—mahogany, black, red, and gold. Net weight of stand, 1½ lbs.

Weight, 4¾ lbs.   24 in case.   Size of case, 6½ cubic feet.

List Price, $4.50 each

The Little Wonder Screw Driver

A fine little tool made exactly like other high grade screw drivers, except for size. Suitable for mechanics. Not a toy.

Very useful for all light work on clocks, telephones, locks, electric fixtures, type-writers, cabinet hardware, radios, etc.

Red stained handles, long steel ferrule. Tempered steel blades, $\frac{1}{8}$ inch in diameter.

Blades are nickel plated so they will not rust or tarnish while on display.

No. 42 Assortment

Comprising four drivers of each size, $1\frac{1}{4}$ inch, 2 inch and 3 inch, mounted on a very attractive red and gold display card. Each card packed in container box.

No. 41. Little Wonder. Size, inches,	$1\frac{1}{4}$	2	3
Length over all, inches,	3	$3\frac{3}{4}$	$4\frac{3}{4}$
List Price, per gross,	**$14.40**	**14.40**	**14.40**

Packed 1 dozen in a box.   60 dozen in case.

No. 42.   Little Wonder, assortment on card as shown.

List Price, $1.20 each

60 Assortments in case.   Weight per dozen assortments, $6\frac{3}{4}$ lbs.

Gross weight, 52 lbs.   Volume, 4 cubic feet.

Tom Thumb Midget Screw Driver

A reliable small screw driver made in three sizes. Very useful in any household or mechanic's kit.

Size,	1	1½	2 inches
Length over all,	3½	4	4½ "

Red fluted handle. Blade, ⁵⁄₃₂ inch in diameter. Forged, hardened and tempered. Ferrule held with bead lock. Nickel plated blade and ferrule.

Just Right in Size for adjusting small screws on Clocks, Sewing Machines, Locks, Typewriters, Door Knobs, Escutcheons, Name Plates, Drawer Pulls, Telephones, Radios, etc.

Silent Salesman Display Card

No. 22 Assortment

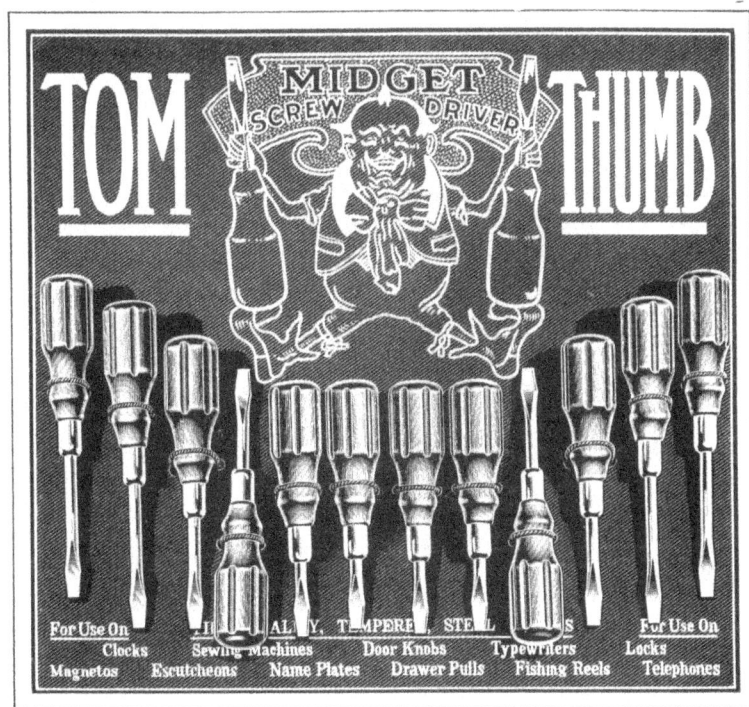

No. 21. Packed 1 dozen in box.   **List Price, $1.80 dozen**
No. 22. Packed 1 dozen on display card. Assorted sizes. **List Price, 1.80 dozen**

5 gross in a case.

The Oh-Kay Screw Driver

A Strong, Reliable Tool at a Popular Price—25 Cents

Description. Fluted hardwood handle—stained dark red and varnished. Nickel plated ferrule.

High grade steel blade, blued finish, with flat sides of point polished bright. The blade of every OH-KAY Screw Driver is tested and the ferrule is held by a new method recently patented by us.

It will not loosen or turn and the blade cannot be turned in the handle.

No. 55 Assortment

The **No. 55** Assortment shown above comprises TWO DOZEN OH-KAY Screw Drivers as follows: 8—4 inch. 8—5 inch. 8—6 inch.

Packed with an attractive heavy cardboard stand—which displays one dozen at a time—each display stand in a separate container box.

The container box holds the extra dozen and as the tools on the display stand are sold the rack is kept filled from this reserve stock.

No. 55 Assortment.   2 dozen with display stand.   **List Price, $6.00 each**

Weight, 7 lbs. each.   Packed 12 Assortments in case.

Eagle Screw Driver

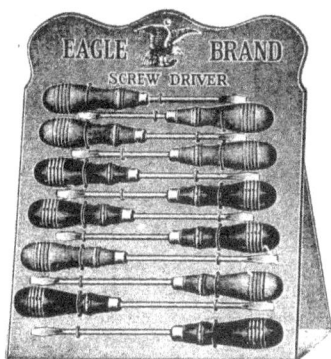

Forged and tempered blades. Bright finish. Diameter, ¼ inch on all sizes. Stained hardwood handle, red or black. A good low-priced Driver for household use. Made in three sizes only. Packed one dozen in box or in assortments.

No. 8 Assortment as shown comprises four 6 inch, four 5 inch and four 4 inch, mounted on a heavy red card that can be stood on a counter or hung up.

No. 9 Assortment comprises the same goods packed 1 dozen in box without display cards.

No. 19 Eagle Screw Drivers — Packed one dozen in box.

Size, inches,	4	5	6
List Price, per doz.,	$1.50	2.00	2.50

No. 8 Assortment

No. 8 Assortment as described above, **$2.00 each.**

No. 9 Assortment as described above, **1.80 each.**

Nos. 8, 9 and **19** packed 24 dozen in case.

Hickory Special Assortment
No. 65

A reliable, attractive screw-driver of a good quality at a popular price. Display stand substantially made from heavy jute board printed to imitate the grain of wood.

Screw Drivers have round fully polished blades securely fastened in handle and will not turn. Points are well hardened and tempered. Dull black fluted handles.

Nickel plated ferrule firmly held to handle by bead lock.

Assortment comprises 4—4 inch, 4—5 inch and 4—6 inch, with display stand. Each in separate container box.

Size, inches,	4	5	6
Diameter, inch,	¼	9/32	5/16

List Price, $3.00 each

Weight, 3¼ lbs. each.

Packed, 24 Assortments in case.

Navy Screw Driver

No. 4001

A Very Serviceable Tool

Round forged blade—hardened, tempered and polished. Heavy steel, nickel plated ferrule with reinforced collar. Fluted hardwood handle, stained red and varnished. A steel rivet through ferrule, handle and blade holds them securely together.

Conforms to U. S. Navy Specification Number 41-S-27

Length of blade, inches,	2	3	4	5	6	8	10	12
Diameter of blade, inch,	$\frac{3}{16}$	$\frac{1}{4}$	$\frac{1}{4}$	$\frac{5}{16}$	$\frac{5}{16}$	$\frac{3}{8}$	$\frac{3}{8}$	$\frac{3}{8}$
List Price, per dozen,	$2.00	2.50	3.00	3.50	4.00	5.50	7.00	9.00

All sizes packed ½ dozen in a box.

The large sizes of No. 4001 Navy Drivers have large double grip handles with extra heavy bull neck ferrules as shown below.

Length of blade, inches,	15	18	24
Diameter of blade, inch,	$\frac{1}{2}$	$\frac{1}{2}$	$\frac{1}{2}$
List Price, per dozen,	$15.00	18.00	20.00

Trojan Screw Driver

No. 209

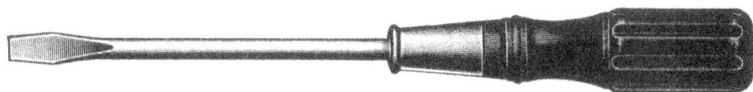

A strong, durable screw driver at a very moderate price. Round polished blade, long steel ferrule, with bead lock. Dull black, fluted handle. Blades are fastened securely into handle.

Length of blade, inches,	2	3	4	5	6	8
List Price, per dozen,	**$1.50**	**2.00**	**2.50**	**3.00**	**3.50**	**4.75**

Packed ½ dozen in box.

Trojan Screw Driver

No. 861

Designed Especially for Automobile Use and Heavy Work

Square forged blade, full polished finish. Points hardened and tempered. Heavy, steel ferrule, nickel plated. Fluted hardwood handle, dull black finish. Blade, handle and ferrule are securely riveted together by a strong steel pin. A wrench or pliers can be applied to the square blade where great power is required.

Size, inches,	3	4	5	6	8
Diameter, inch,	$\frac{1}{4}$	$\frac{1}{4}$	$\frac{5}{16}$	$\frac{5}{16}$	$\frac{5}{16}$
List Price, per dozen,	**$2.50**	**3.00**	**3.50**	**4.00**	**5.50**

Packed ½ dozen in box.

Trojan Bull Dog

No. 862

A Short, Heavy Tool for Hard Jobs where Space is Limited

This tool has a short, square blade 2 x $\frac{5}{16}$ inches and measures only $6\frac{1}{4}$ inches over all. The handle is large in diameter so heavy screws can be started easily.

Dull black handle, nickel plated ferrule. Full polished blade. Handle, ferrule and blade securely riveted together.

List Price, $3.00 per dozen

Packed $\frac{1}{2}$ dozen in box.   24 dozen in case.

Standard Screw Driver

No. 460

A Very Good Popular Priced Driver

Round, polished steel blade, nickel plated ferrule with bead lock. Red stained handles, popular fluted design. Blades are fastened into handles very securely. They will not turn and the points are well hardened and tempered.

Length of blade, inches,	2	2½	3	4	5	6	8
Diameter, inch,	$\frac{3}{16}$	$\frac{3}{16}$	$\frac{7}{32}$	$\frac{1}{4}$	$\frac{9}{32}$	$\frac{5}{16}$	$\frac{5}{16}$
List Price, per dozen,	**$2.00**	**2.25**	**2.50**	**3.00**	**3.50**	**4.00**	**5.50**

Packed 1 dozen in box.

Service Screw Driver

No. 2530

A very low priced but serviceable driver for household use. Red stained fluted handle, nickel plated brass ferrule. Round blades, bright finish.

Length of blade, inches,		2	3	4	5	6	8
Diameter, inch,		3_{16}	1_4	1_4	1_4	5_{16}	5_{16}
List Price, per dozen,		$1.50	2.00	2.50	3.00	3.50	4.75

Packed 1 dozen in box.

Bridgeport Screw Driver

No. 44

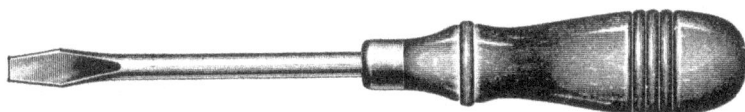

A Strong, Durable, Low Priced Driver

Round forged, polished blade. Hardened and tempered point. Hardwood handles finished in bright cherry red. Nickel plated ferrules, held firmly by bead lock. Each tool branded.

Size, inches,		2	3	4	5	6	8	10	12
List Price, per dozen,		$1.50	2.00	2.50	3.00	3.50	4.75	6.00	8.00

Packed 2 and 3 in., 1 dozen in a box. All other sizes ½ dozen in a box.

Dwarf Screw Driver Assortment
No. 32

Comprising twelve Midget Screw Drivers of the following sizes:

4–1¼ inch   4–2 inch   4–3 inch

Blades ⅛ inch in diameter, red handle, steel ferrule.   Display card in green and white.

60 Assortments in a case.   Weight, 5¼ lbs. per dozen Assortments.

List Price, $1.20 each

Reliance Screw Driver Set
No. 17

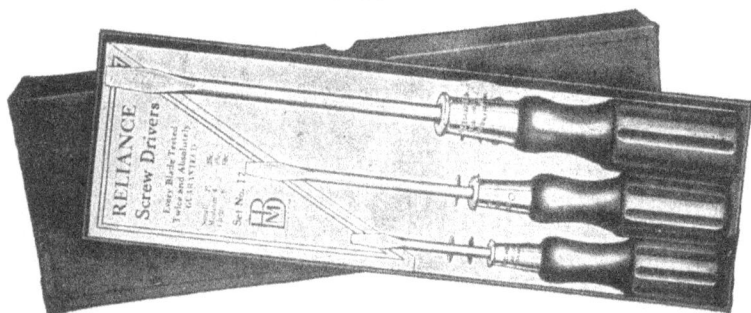

Comprising the three most popular sizes of Reliance Screw Drivers:

1—2½ inch   1—4 inch   1—6 inch

Packed in nicely finished black box, lined with red.

Screw Drivers of the highest quality.

Every One Warranted

Weight per dozen sets 10½ lbs.   36 sets to a case.

List Price, $1.00 per set

Big Bully Screw Driver

And "Odd Job" Tool

No. 300

An Exceptionally Strong, Powerful Tool—Serves as a Pry Bar, Chisel, Heavy Duty Screw Driver, Etc.

Especially adapted for automotive work. Note the various uses. A tool needed in every workshop, car and garage.

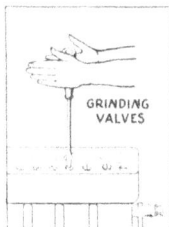

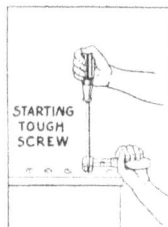

The blade of the Big Bully is $\frac{7}{16}$ inch octagonal tool steel tempered from end to end and it extends clear through the handle. The handle is large in diameter so that large screws can be turned easily. Ferrule, handle and blade are securely riveted together.

The Big Bully is attractively finished with polished mahoganized handle, nickel plated ferrule and black blade polished on flat sides of point.

One size only—8 x $\frac{7}{16}$ inch blade.   **List Price, $9.00 per dozen**

Packed $\frac{1}{2}$ dozen in box.   Weight, 10 lbs. per dozen.   12 dozen in case.

Pribar Screw Driver

No. 242

An Exceptionally Strong Tool for Heavy Work either as
a Screw Driver, Tire Tool or Pry Bar

Especially suitable for motor car repair work. The blade is made of octagonal tool steel 7/16 inch in diameter, tempered from end to end. As the name indicates, it is made to "stand up" and not bend out of shape when used as a lever, prying off tires, etc.

Blade 7/16 inch x 8 inch. Natural steel finish with sides of point polished. Handle— black rubberoid finish. Ferrule—heavy steel, unpolished. Ferrule, blade and handle fastened together with heavy steel rivet.

List Price, $9.00 per dozen

Packed 1/2 dozen in box.   18 dozen in case.

B. M. Co. Screw Driver

No. 112

Flat forged polished blade. Flat varnished beech handle, natural color. Brass ferrule. Tempered point.

Size, inches,	2	3	4	5	6	8	10	12
List Price, per dozen,	$1.50	2.00	2.50	3.00	3.50	4.75	6.00	8.00

Two and three inch packed 1 dozen in box. All other sizes 1/2 dozen in box.

Rex Screw Driver
No. 29

For Ford Magneto Posts, Motorcycles, Etc.

A sturdy small screw driver for light automotive work. Measures only 5 inches over all and can be used in places where space is limited. Conveniently carried in the pocket. Blade full polished, 1/4 by 1 1/2 inches. Dull black fluted handle.

Packed 1 dozen in box. **List Price, $2.00 per dozen**

Auto Driver
No. 221

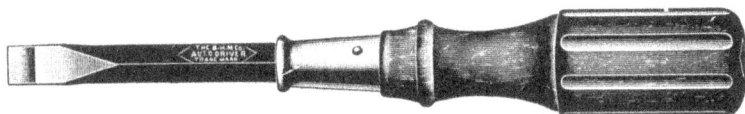

Designed especially for automotive work having a square forged shank permitting the use of a wrench or pliers in starting large or rusty screws, etc. Blade diameter $5/16$ inch and extends clear through the handle. Oil tempered point.

Fluted hardwood handle, heavy bright steel ferrule. Blade, ferrule and handle are securely riveted together. Handle is stained weathered oak color and varnished.

Black oil finished blade with flat sides of point polished, made in one size only.

No. 221. 3 inch Auto Driver. **List Price, $3.00 per dozen**

Packed 1/2 dozen in box.

Acme Screw Driver
No. 52

Ideal for Sewing Machines, Etc.

1 1/2 inch round steel blade hardened and tempered. Black enameled handle. Nickel plated ferrule. 1 dozen in a box.

List Price, $15.00 per gross

Screw Drivers

TABLE OF WEIGHTS AND CUBIC MEASUREMENTS PER DOZEN ON ALL SCREW DRIVERS SHOWN ON PRECEDING PAGES

Page No.												
21	**No. 92 Matchless Screw Driver**	**Size**	**2**	**2½**	**3**	**4**	**5**	**6**	**8**	**10**	**12**	**inch**
		Legal weight per dozen	1⅜	1½	2¼	2½	3⅞	4¼	7	7⅝	8¼	pounds
		Cubic Measurement per doz	77	77	128	128	160	178	250	288	330	cubic ins.
21	**No. 93 Matchless Machinist's Screw Driver**	**Size**	...	...	**3**	**4**	**5**	**6**	**8**	**10**	**12**	**inch**
		Weight per dozen	...	...	3¼	3½	5½	6	7½	10	11¼	pounds
		Cubic Measurement per doz	...	...	125	125	161	178	244	276	316	cubic ins.
21	**No. 94 Matchless Mechanic's Screw Driver**	**Size**	**2**	**2½**	**3**	**4**	**5**	**6**	**8**	**10**	**12**	**inch**
		Weight per dozen	1⅝	1¾	2¾	3¼	4¾	5¼	8	8¾	9½	pounds
		Cubic Measurement per doz	80	90	92	125	144	178	251	288	330	cubic ins.
22	**No. 95 Matchless Electrician's Screw Driver**	**Size**	...	**2½**	**3**	**4**	**5**	**6**	**8**	**10**	**12**	**inch**
		Weight per dozen	...	1¾	2¾	3¼	4¾	5¼	8	8¾	9½	pounds
		Cubic Measurement per doz	...	90	92	125	144	178	251	288	330	cubic ins.
22	**No. 96 Matchless Cabinet Screw Driver**	**Size**	...	**2½**	**3½**	**4½**	**5½**	**6½**	**8½**	**10½**	**12½**	**inch**
		Weight per dozen	...	1¾	1⅞	2	2⅛	2¼	2½	2¾	3	pounds
		Cubic Measurement per doz	...	90	101	101	113	113	165	165	187	cubic ins.
22	**No. 97 Matchless Heavy Machinist's Screw Driver**	**Size**	...	**2½**	**3½**	**4½**	**5½**	...	...	...	...	**inch**
		Weight per dozen	...	5¼	5¾	8¼	8¾	...	...	...	...	pounds
		Cubic Measurement per doz	...	169	169	229	229	...	...	...	...	cubic ins.
23	**No. 98 Matchless Electrician's Screw Driver**	**Size**	...	**2½**	**3½**	**4½**	**5½**	**6½**	**8½**	**10½**	**12½**	**inch**
		Weight per dozen	...	1½	1⅝	1¾	1⅞	2	2¼	2½	2¾	pounds
		Cubic Measurement per doz	...	90	101	101	113	113	165	165	187	cubic ins.
23	**No. 99 Baby Matchless Screw Driver,** Weight per dozen, 1 pound; Cubic Measurement per dozen, 45 cubic inches.											
23	**No. 970 Matchless Little Bully**	Weight per dozen, 4½ pounds; Cubic Measurement per dozen, 165 cubic inches.										

Page No.												
24	**No. 100 Matchless**	**Shock Proof Screw Driver**	...	**2½**	**3½**	**4½**	**5½**	**6½**	**8½**	**10½**	**12½**	**inch**
		Legal weight per dozen	...	1¾	1⅞	2	2⅛	2¼	2⅜	2⅝	3	pounds
		Cubic Measurement per doz	...	80	95	100	120	120	165	185	210	cubic ins.
24	**No. 60 Asst. Matchless Shock Proof,** Legal weight each, 2¾ pounds; Cubic Measurement each, 170 cubic inches.											
25	**No. 32 Challenge Mechanic's Screw Driver**	**Size**	**2**	**2½**	**3**	**4**	**5**	**6**	**8**	**10**	**12**	**inch**
		Legal weight per dozen	1¼	1½	2	2¼	4	4¼	7¼	8¼	8⅞	pounds
		Cubic Measurement per doz	90	92	106	118	160	164	270	367	404	cubic ins.
25	**No. 36 Challenge Electrician's Screw Driver**	**Size**	...	**2½**	**3½**	**4½**	**5½**	**6½**	**8½**	**10½**	**12½**	**inch**
		Legal weight per dozen	...	1⅜	1½	1⅝	1¾	1⅞	2⅛	2⅜	2⅝	pounds
		Cubic Measurement per doz	...	91	91	101	124	124	146	169	169	cubic ins.
25	**No. 37 Baby Challenge Screw Driver,** Weight per dozen, ¾ pounds; Cubic Measurement per dozen, 45 cubic inches.											
26	**No. 38 Giant Challenge Screw Driver**	**Size**	**15**	**18**	**24**	**inch**						
		Legal weight per dozen	19½	21½	27	pounds						
		Cubic Measurement per doz	570	650	785	cubic ins.						
26	**No. 252 Vulcan Screw Driver**	**Size**	...	**2**	**3**	**4**	**5**	**6**	**8**	**10**	**12**	**inch**
		Legal weight per dozen	...	1⅜	2¼	2⅝	4⅛	4⅜	5½	6¼	7¼	pounds
		Cubic Measurement per doz	...	75	128	128	157	171	260	276	292	cubic ins.
26	**No. 254 Vulcan Cabinet Screw Driver**	**Size**	...	...	...	**2½**	**3½**	**4½**	**5½**	**6½**	**8½**	**inch**
		Legal weight per dozen	...	...	...	1⅜	1½	1⅝	1¾	1⅞	2⅛	pounds
		Cubic Measurement per doz	...	...	...	91	91	101	124	124	146	cubic ins.
27	**No. 102 Hercules Screw Driver**	**Size**	**2**	**3**	**4**	**5**	**6**	**7**	**8**	**10**	**12**	**inch**
		Legal weight per dozen	1¾	2	4⅛	6⅜	6⅝	6⅞	10	11¼	11¾	pounds
		Cubic Measurement per doz	76	88	110	145	158	170	232	266	291	cubic ins.
27	**No. 12 Hercules Heavy Screw Driver,** Weight per dozen, 9 pounds; Cubic Measurement per dozen, 158 cubic inches.											
28	**No. 40 Assortment Matchless No. 94 Screw Driver,** Weight each, 5½ pounds; Cubic Measurement each, 310 cubic inches.											

Page No.														
29	**No. 7 Reliance Screw Driver**	**Size**			**2**	**2½**	**3**	**4**	**5**	**6**	**8**	**10**	**12**	**inch**
		Legal weight per dozen			1½	1¾	2¼	2¼	4	4¼	6	7⅜	8¾	pounds
		Cubic Measurement per doz			80	80	128	128	150	156	260	276	292	cubic ins.
29	**No. 35 Assortment Reliance Screw Driver,** Weight each, 4½ pounds; Cubic Measurement each, 345 cubic inches.													
30	**No. 42 Asst. Little Wonder,** Legal weight per doz. asst., 6¾ pounds; Cubic Measurement per doz. asst., 1056 cubic inches.													
30	**No. 41 Any Size Little Wonder,** Legal weight per gross, 3½ pounds; Cubic Measurement per gross, 200 cubic inches.													
31	**No. 22 Asst. Tom Thumb,** Legal weight per doz. asst., 10½ pounds; Cubic Measurement per doz. asst., 1320 cubic inches.													
32	**No. 55 Assortment Oh-Kay,** Legal weight each, 7¼ pounds; Cubic Measurement each, 338 cubic inches.													
33	**No. 8 Eagle Assortment Screw Drivers,** Weight each, 3 pounds; Cubic Measurement each, 412 cubic inches.													
33	**No. 65 Hickory Special Assortment,** Weight each, 3¼ pounds; Cubic Measurement each, 410 cubic inches.													
34	**No. 4001 Navy Screw Drivers**	**Size**	**2**	**3**	**4**	**5**	**6**	**8**	**10**	**12**	**15**	**18**	**24**	**inch**
		Legal weight per dozen	1¼	2	2½	3⅞	4⅛	6	7⅜	8¾	14½	21½	27	pounds
		Cubic Measurement per doz	80	88	121	156	168	214	255	285	570	650	785	cubic ins.
35	**No. 209 Trojan Screw Driver**	**Size**			...	...	...	**2**	**3**	**4**	**5**	**6**	**8**	**inch**
		Legal weight per dozen			...	...	...	⅞	1½	2¼	2½	4	4½	pounds
		Cubic Measurement per doz			...	...	...	35	80	128	142	175	203	cubic ins.
35	**No. 861 Trojan Screw Driver**	**Size**			...	...	...	...	**3**	**4**	**5**	**6**	**8**	**inch**
		Legal weight per dozen			...	...	...	...	2	2¾	4½	5	6	pounds
		Cubic Measurement per doz			...	...	...	...	100	118	147	189	214	cubic ins.
36	**No. 862 Trojan Bull Dog,** Legal weight per dozen, 3¾ pounds; Cubic Measurement per dozen, 172 cubic inches.													
36	**No. 460 Standard Screw Driver**	**Size**			...	...	**2**	**2½**	**3**	**4**	**5**	**6**	**8**	**inch**
		Legal weight per dozen			...	...	⅞	1	1½	2¼	2½	4	4½	pounds
		Cubic Measurement per doz			...	...	35	60	80	128	142	175	203	cubic ins.

Page No.												
37	**No. 2530 Service Screw Driver**	**Size**				**2**	**3**	**4**	**5**	**6**	**8**	**inch**
		Legal weight per dozen				½	1½	1¾	2	3	3½	pounds
		Cubic Measurement per doz				48	66	98	116	138	150	cubic ins.
37	**No. 44 Bridgeport Screw Driver**	**Size**	**2**	**3**	**4**	**5**	**6**	**8**	**10**	**12**		**inch**
		Legal weight per dozen	⅞	1⅝	2	3½	3¾	6⅛	7⅛	8¼		pounds
		Cubic Measurement per doz	49	84	115	152	162	212	286	292		cubic ins.

38 **No. 32 Dwarf Assortment,** Legal weight per doz. assts., 5¼ pounds; Cubic Measurement per doz. assts., 660 cubic inches.

38 **No. 17 Reliance Screw Driver Set,** Legal weight per doz. sets, 10 pounds; Cubic Measurement per doz. sets, 860 cubic inches.

39 **No. 300 Big Bully Screw Driver,** Legal weight per dozen, 10 pounds; Cubic Measurement per dozen, 310 cubic inches.

40 **No. 242 Pribar 8 inch,** Legal weight per dozen, 7½ pounds; Cubic Measurement, 174 cubic inches.

Page No.											
40	**No. 112 B. M. Co. Screw Driver**	**Size**	**2**	**3**	**4**	**5**	**6**	**8**	**10**	**12**	**inch**
		Legal weight per dozen	¾	1¼	2⅛	2⅞	3¼	5¾	7	8	pounds
		Cubic Measurement per doz	50	84	121	160	168	285	323	360	cubic ins.

41 **No. 29 Rex Screw Driver,** Weight per dozen, 1 pound; Cubic Measurement per dozen, 47 cubic inches.

41 **No. 221 Autodriver 3 inch,** Weight per dozen, 4½ pounds; Cubic Measurement, 146 cubic inches.

41 **No. 52 Acme Sewing Machine Driver,** Weight per gross, 6½ pounds; Cubic Measurement per gross, 492 cubic inches.

Fay-O-Rite Valve Lifter
No. 1

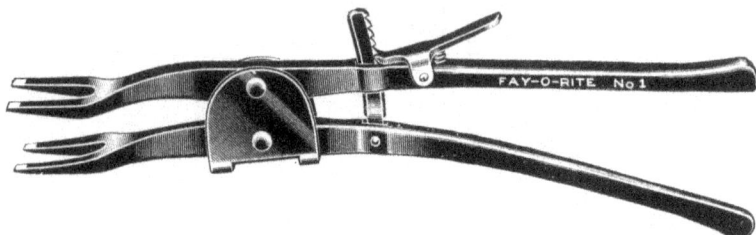

A FORGED STEEL TOOL OF FINEST QUALITY
FOR REMOVING VALVES FROM ALL TYPES OF GAS ENGINES

The Fay-O-Rite was designed by a practical mechanic who knew from experience the faults of other lifters and eliminated them in this new tool.

It should not be confused with cheap cast iron affairs, similar in shape. **Note the difference.**

LIFTS STRAIGHT

WON'T SLIP

SAFETY FIRST—IT WON'T SLIP

When in use the forked ends operate at an angle which holds them in place under tension. They will not slip out— a common fault with Valve Lifters—and it lifts straight up. The cut at the left illustrates this.

QUICK ACTING

One movement compresses the valve spring so key can be removed. The ratchet lock holds the jaws in any position and is absolutely positive. Sharp, deep teeth eliminate slipping and the spring lock will never release accidentally.

ROCKER LINKS

ADJUSTS ITSELF

SELF-ADJUSTING

The two points are joined by "Rocker Links," a patented feature, which permits the forked ends to work slightly forward or back, thus adjusting themselves when pressure is applied. This feature prevents cramping or binding of valve stem or washer. No more bent valve stems.

CHISEL POINTS

When forked ends are together as shown, they form a shape like a chisel so they can be easily inserted under the spring. This is a very important exclusive feature of the Fay-O-Rite.

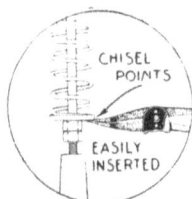

CHISEL POINTS

EASILY INSERTED

DURABLE

All parts of the Fay-O-Rite are made of high grade forged steel, and the construction is so simple and strong that it will "Stand Up" under the hardest use for many years. Every part guaranteed.

ADAPTABLE

Forked ends of the Fay-O-Rite are especially strong and tough. They can be bent in a vise into any desired shape, and they will not crack nor break. This permits using the Fay-O-Rite for jobs on old types of motors where no other could be used.

Length, 12½ inches.   Weight, 25 ounces.

6 dozen in case.   **List Price, $18.00 per dozen**

Gross weight of case, 119 lbs.   Net weight, 113 lbs.

Volume, 4½ cubic feet.

The Matchless Tire Tool

A Pry Bar of Exceptional Strength for Use on Automobile Tires

The **Matchless Tire Tool** is made from one piece of spring tempered steel, forged to taper from end to end, and slightly curved at the point to just the right shape for tire work.   Needed in every car, garage, or repair shop.

Two hardwood scales are riveted securely to this bar forming an unbreakable oval handle.   The wooden portions are light brown in color and oil finished.

Length, 11½ inches.   Weight, 19 ounces.

No. 38.   Natural steel finish with polished end.   **$9.00 dozen net**

No. 39.   Full polished all over.   **12.00 dozen net**

⅓ dozen in box.   6 dozen in case.

Gross weight, 108 lbs.   Size of case, ¾ cubic foot.   Net weight, 90 lbs.

Hi-Power Tire Tool

A tire tool of fine quality and of perfect design at a very moderate price.   The use of tough steel tempered the full length gives great strength and it will stand an enormous prying strain.   One end slightly curved at the proper angle to hook over rims or under tire casings.   End of handle semi-knurled for a good grasp and to prevent slipping when hands are greasy.

No. 275.   Bright finish.   **List Price, $3.00 per dozen**

No. 280.   Black, rubberoid handle, nickel plated and highly buffed.

                                                     List Price, $6.00 per dozen

Weight, 22 ounces.   Packed 1 dozen in box.   Length, 18 inches.

12 dozen in a case.   Gross weight, 225 lbs.

Rex Tire Tool
No. 154

Size, ¼ x 1⅛ x 11¾ inches.

A very strong, small tire tool for motor car or motorcycle tires.   Made of spring tempered steel, black oil finish with both ends polished and blued.

Weight, 13 ounces each.   Length, 11¾ inches.

List Price, $6.00 per dozen

Packed ½ dozen in box.   12 dozen in case.

Tire Tool

No. 1154

For Light Auto Tires or Motorcycles

Size, $\frac{1}{4}$ x $1\frac{1}{8}$ x $11\frac{3}{4}$ inches.

Made of good steel and tempered from end to end, bright finish.

List Price, $4.00 per dozen

Weight, 13 ounces.   Length, $11\frac{3}{4}$ inches.

Packed $\frac{1}{2}$ dozen in box.   18 dozen in case.

Thor Tire Tool

No. 250

A Very Strong Tool for Heavy Work

Size, $\frac{3}{8}$ x $\frac{7}{8}$ x 16 inches.

Made of very tough steel and tempered from end to end. Enormous power can be applied with the Thor and it will stand the hardest prying strains. One end is formed in a straight taper while the other is slightly curved at the proper angle to hook over rims or under tire casings. This tool is a favorite with tire repairmen.

List Price, $4.80 per dozen

Weight, 24 ounces.   Length, 16 inches.

Packed $\frac{1}{2}$ dozen in box.   12 dozen in case.

Gross weight, 227 lbs.

Master Auto Hammer

No. 185

A Strictly High-Grade Tool for Automotive Work

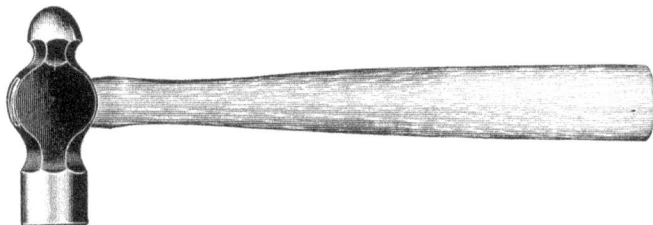

Every One Guaranteed

The Master Auto Hammer is just right in size and weight for automobile use. Head is made of forged steel of best quality. Short, first quality white hickory handle.

Weight of head, 13 ounces.   Length over all, 11 inches.

List Price, $9.00 per dozen

Packed ½ dozen in box.   10 dozen in case.

Auto Hammer

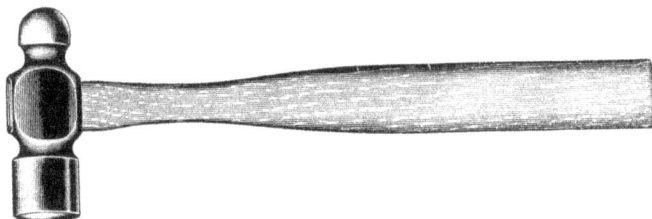

Head made of high grade, drop forged steel, properly hardened, tough hickory handle stained brown.

No. **18**—Natural steel finish with polished hammer head and pein.

List Price, $6.00 per dozen

No. **8**—Black rubberoid finish on head with polished face.

List Price, $5.00 per dozen

12 ounce head.   Length over all, 11 inches.

Packed ½ dozen in box.   10 dozen in case.

Size of case, 2½ cubic feet.

Gross weight, 140 lbs.   Net weight, 115 lbs.

Matchless Adjustable Hack Saw

No. 30

A Superior Tool

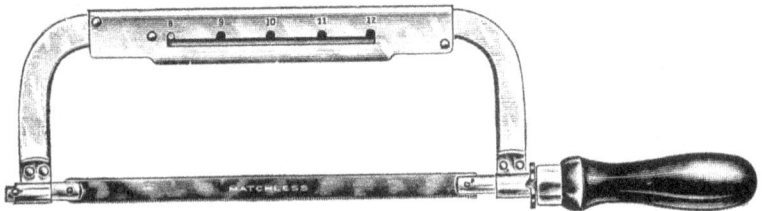

Every requirement of a high grade saw is combined in the Matchless. The rigid frame, speedy adjustment and extra strong construction of this tool are quickly appreciated by mechanics.

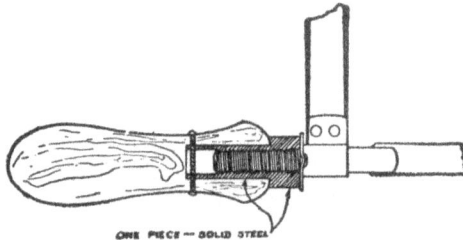

ONE PIECE — SOLID STEEL

Great strength and durability in the handle and blade stretcher assembly are secured by the construction shown above. All metal parts are highly nickel plated and buffed. Black rubberoid finished handle made of hardwood.

One 8 inch blade with each frame.

Length, 15 inches.   Weight, 17 ounces.

List Price, $24.00 per dozen

Packed one in a box.   6 dozen in a case.
Gross weight, 122 lbs.   Size of case, $6\frac{1}{2}$ cubic feet.   Net weight, 78 lbs.

The Hy-Power Side Cutting Plier

A Tool of Wonderful Strength and Durability for All Kinds of Jobs

No. 320

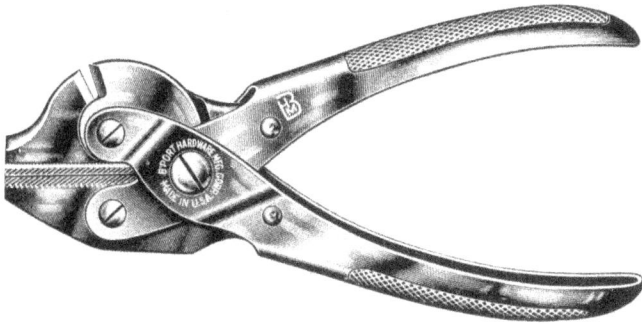

Ideal for the "Handy-man" who makes his own repairs, as well as for mechanics who require the very best in tools.

Compound leverage and parallel jaws give astonishing cutting power and a vise-like grip.

A pressure of 50 lbs. on the handles exerts a tremendous pressure of 1200 lbs. between the cutting edges and a squeeze of 300 lbs. between the jaws.

Hollow Box Joint keeps the jaws and cutting edges always in line, and a wire or rod can be thrust clear through. The jaws grip it rigidly at any point. Fine for stretching or bending wire.

Side Cutter located at a point where it will work at any angle and cut in places impossible to reach with ordinary pliers.

Interchangeable Jaws forged from special steel—can be easily replaced when worn out by hard shop use.

No. 320. Full nickel plated, beautifully finished. **List Price, $42.00 per dozen**

No. 210. Embodying all of the good features of the No. 320 at a moderate price. Ends of Jaws polished, mottled handles. **List Price, $21.00 per dozen**

Size, 7 inches.

Weight, 11 ounces each. Packed in individual boxes.
6 dozen in a case. Size of case, 1½ cubic feet.
Gross weight, 70 lbs. Net weight, 48 lbs.

Matchless Combination Pliers

With Alligator Jaws and Slip Joint

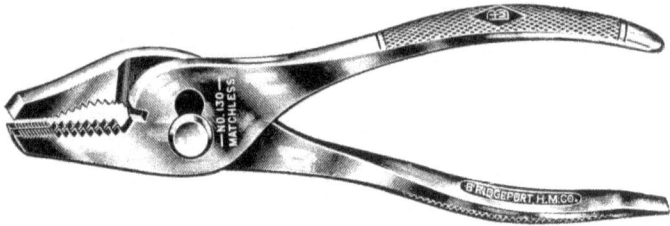

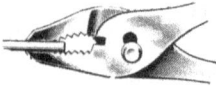

The Most Practical Plier of all for a Multitude of Uses in the Workshop, Home, Motor Car, Garage or Farm

Size, 7 inches

Alligator Jaws. Grip objects of any shape and hold them rigidly.

Slip Joint. Jaws open very wide and the curved upper jaw gives a pipe wrench grip on objects from $\frac{3}{8}$ to $1\frac{1}{4}$ inches in diameter.

Slender Nose. Beveled at the ends—small parts can be reached and adjusted in difficult places where an ordinary plier is useless.

Screw Driver. A screw driver point on one handle for emergency use.

Wire Cutter and Holder. A practical wire cutter between the jaws. Groove in the lower jaw holds wire or other small round parts against the teeth of upper jaws giving a vise-like grip.

Made of forged steel of finest quality with milled teeth, knurled handle and cutters carefully hardened.

No. 130. 7″ Beautifully polished and full nickel plated. List Price, $12.00 per dozen
No. 133. 7″ Blued finish with polished jaws. List Price, 9.00 per dozen
No. 1133. 7″ Natural steel finish with polished nose. List Price, 6.00 per dozen

Packed ½ dozen in a box. 12 dozen in a case.

Gross weight, 80 lbs. Net weight, 60 lbs.

Matchless Combination Pliers
No. 118

A Very Popular Pattern for General Use

Combines flat nose plier with gas burner grip and side cutters, two shear wire cutters, reamer and screw driver.

Made of high grade steel, drop forged from bar with jaws and cutting edges carefully hardened and tempered.

No. 118. Blued handles, nose polished on both sides and edges as illustrated above.

Size, inches,	$5\frac{1}{2}$	6	7	8
List Price, per dozen,	$11.75	12.70	14.70	16.80

No. 1180. Same as No. 118 but with jaws polished on one side only and with black oil finish.

Size, inches,	$5\frac{1}{2}$	6	7	8
List Price, per dozen,	$11.00	12.00	14.00	16.00

No. 236. Semi-polished handles, highly polished jaws.

Size, inches,	$5\frac{1}{2}$	6	7	8
List Price, per dozen,	$13.75	15.00	17.00	19.00

No. 632

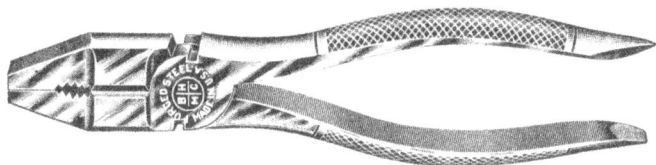

No. 632. Similar to pliers listed above but very attractively finished—full polished all over.

Size, inches,	$5\frac{1}{2}$	6	7	8
List Price, per dozen,	$14.80	16.00	18.00	21.00
Weight per dozen, lbs., all styles,	$3\frac{1}{2}$	$5\frac{1}{8}$	$6\frac{1}{4}$	$9\frac{1}{4}$

All packed ½ dozen in box—each one tested and warranted.

Matchless Side Cutting Pliers
No. 142

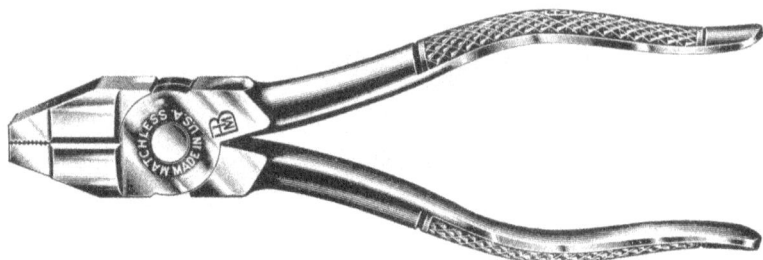

Made of fine quality steel forged from the bar. Cutting edges and jaws carefully heat-treated. Blued steel scored handles, polished head as illustrated.

Made in one size only, 6 inch.   **List Price, $10.80 per dozen**
Packed ½ dozen in box.   Weight, 4⅞ lbs. per dozen

Every one tested and warranted

Matchless End Cutting Nippers
No. 116

For cutting all kinds of soft wire. Correctly designed, well made and very durable. Made entirely of carbon steel, drop forged from the bar. Jaws hardened, oil tempered and tested, natural steel finish with polished jaws as shown in above cut.

Every one tested and warranted

Size, 5½ inches.   Net weight per dozen, 4¾ lbs.   **List Price, $9.00 per dozen**
Size, 6½ inches.   Net weight per dozen, 5¾ lbs.   **List Price, 10.00 per dozen**

Packed ½ dozen in a box.   2 gross in case.

Size of case, 3¼ cubic feet.

Compound Cutting Nippers

No. 113

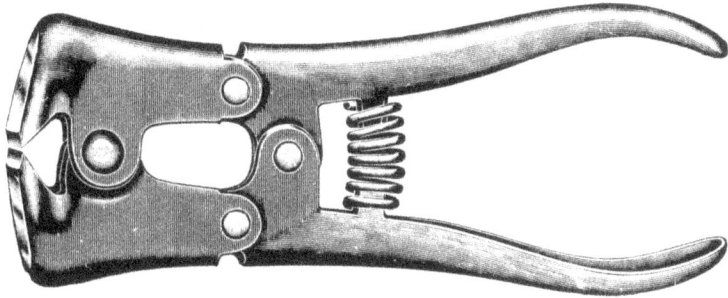

Each one warranted

Forged from fine quality sheet steel carefully hardened and oil tempered. Compound lever insures easy cutting. This nipper will cut any soft wire, but is not intended for cutting piano wire. Natural steel finish, faces of jaws polished bright.

Size, 5½ inches.   **List Price, $9.00 per dozen**
Gross weight, 115 lbs.   Net weight, 100 lbs.   Volume, 2¼ cubic feet.
Size, 6½ inches.   **List Price, $10.00 per dozen**
Gross weight, 120 lbs.   Net weight, 105 lbs.   Volume, 2½ cubic feet.

Both sizes are packed ½ dozen in a box, 2 gross in a case.

Keyes' Wire Cutters

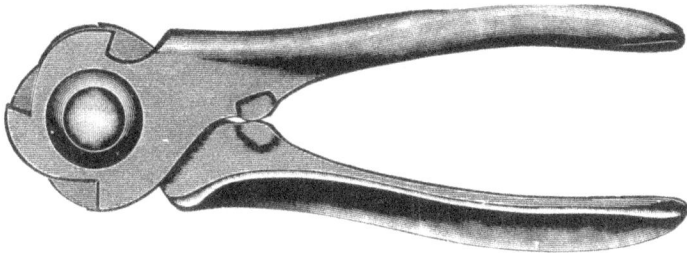

No. 20. Black, 3⅞ inches.   **List Price, $1.50 per dozen**
Packed ½ dozen in box.   2 gross in case.
                  Size of case, ¾ cubic foot.
Gross weight, 62 lbs.   Net weight, 50 lbs.

No. 60. Black, 5 inches.   **List Price, $1.75 per dozen**
Packed ½ dozen in a box.   2 gross in a case.
                  Size of case, 1 cubic foot.
Gross weight, 69 lbs.   Net weight, 54 lbs.

The PARK Flat Nose Pliers
No. 240

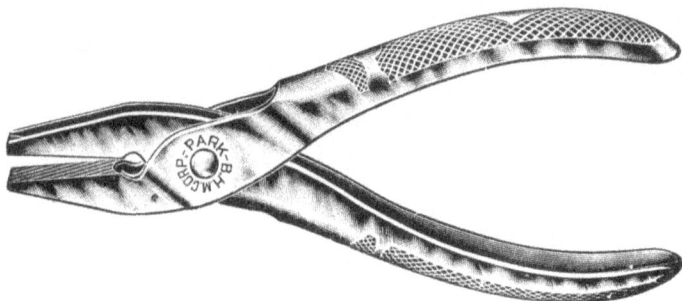

Polished and Nickel Plated Finish
A Reliable Tool—Hardened and Tempered.   Scored Jaws
List Prices:
4 inch, **$3.00** per dozen.   5 inch, **$4.00** per dozen.   6 inch, **$5.00** per dozen.

Packed 1 dozen in a box.   2 gross in a case.

Size of case and weight same as No. 24 Toro Plier shown on next page.

The PARK Cutting Pliers
No. 260

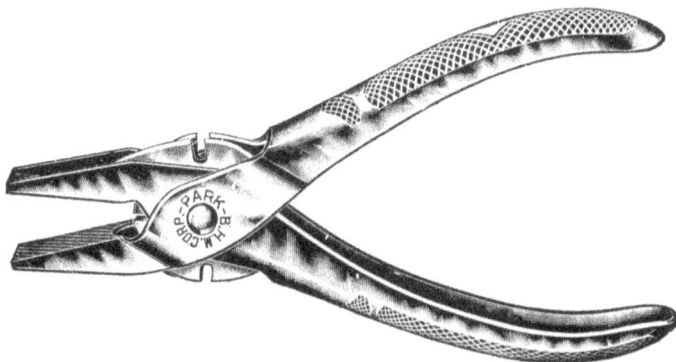

Polished and Nickel Plated Finish
A Reliable Tool—Hardened and Tempered.   Scored Jaws
Extra Wire Cutter between Jaws
List Prices:
5 inch, **$5.00** per dozen.   6 inch, **$7.00** per dozen.

Packed 1 dozen in a box.   2 gross in a case.

Size of case and weight same as No. 26 Toro Pliers shown on next page.

Toro Steel Pliers
No. 24

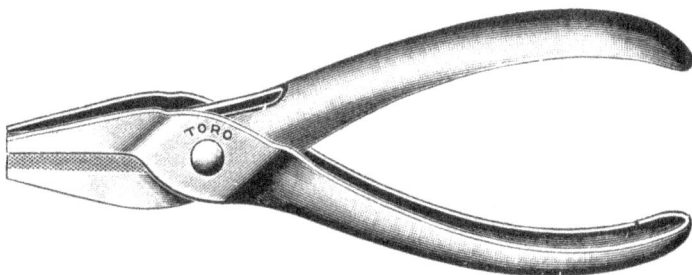

Made of C. R. Steel, Scored Jaws, Bright Polished Finish
List Prices:

4 inch.	5 inch.	6 inch.
$1.00 per dozen.	**$1.50** per dozen.	**$2.00** per dozen.
Size of case, ½ cubic foot.	Size of case, 1¼ cubic feet.	Size of case, 1¾ cubic feet.
Gross weight, 36 lbs.	Gross weight, 52 lbs.	Gross weight, 74 lbs.
Net weight, 24 lbs.	Net weight, 39 lbs.	Net weight, 60 lbs.

Toro Cutting Pliers
No. 26

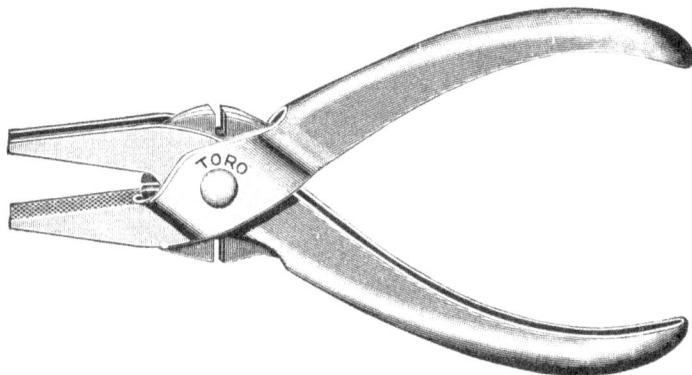

Made of C. R. Sheet Steel, Scored Jaws, Bright Polished Finish
Hardened and Tempered Cutting Disks.   Extra Cutter between Jaws
List Prices:

5 inch.	6 inch.
$2.00 per dozen.	**$3.00** per dozen.
Size of case, 1¼ cubic feet.	Size of case, 1¾ cubic feet.
Gross weight, 73 lbs.	Gross weight, 110 lbs.
Net weight, 60 lbs.	Net weight, 96 lbs.

Nos. 24 and 26 packed 1 dozen in box.   2 gross in case.

Toro Special Pliers

No. 126

No. 124.   Toro Special Pliers, plain nose without cutters. Bright finished knurled handles.

Size, inches,	4	5	6
List Price, per dozen,	$1.00	1.50	2.00

Size of case, 12x10x8 in.	Size of case, 14x14x11 in.	Size of case, 18x14x12 in.
Gross weight, 36 lbs.	Gross weight, 52 lbs.	Gross weight, 74 lbs.
Net weight, 24 lbs.	Net weight, 39 lbs.	Net weight, 60 lbs.

No. 126.   Toro Special Pliers with cutters as shown above. Hardened and tempered discs. Extra cutter between jaws. Bright finished knurled handles.

Size, inches,	5	6
List Price, per dozen,	$2.00	3.00

Size of case, $1\frac{1}{4}$ cubic feet.	Size of case, $1\frac{3}{4}$ cubic feet.
Gross weight, 73 lbs.	Gross weight, 110 lbs.
Net weight, 60 lbs.	Net weight, 96 lbs.

Nos. 124 and 126 packed 1 dozen in box. 2 gross in case.

Ice Pick
No. 3

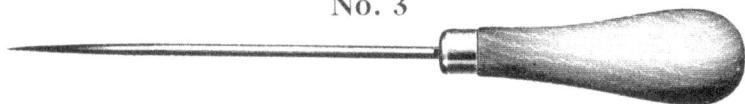

Tempered steel blade—bright finish. Slender point. Natural finished handle.
Diameter 5⁄32 inch. 1 dozen in a box. 3 gross in a case.
Length of blade, 6 inches.
List Price, $18.00 per gross
Gross weight, 75 lbs.   Net weight, 50 lbs.   Volume, 3 cubic feet.

Ice Pick
No. 4

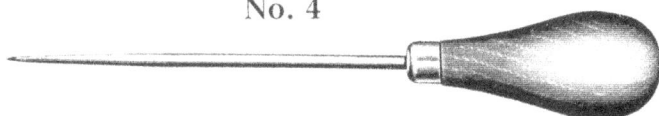

Blade, 5 inches in length. Full tempered. Nickel plated ferrule.
Red stained handle. Diameter, 5⁄32 inch.
Packed 1 dozen in a box.   3 gross in a case.
List Price, $15.00 per gross
Gross weight, 60 lbs.   Net weight, 40 lbs.   Volume, 2½ cubic feet.

Ice Pick
No. 7

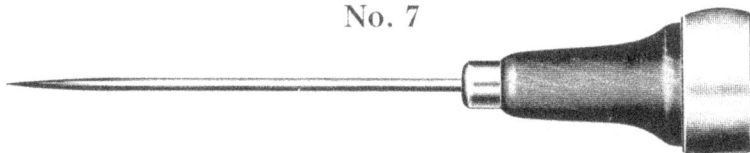

Made of high grade steel—spring tempered. Needle point—polished blade.
Red stained handle. Nickel plated blade, ferrule and metal cap.
Length of blade, 6 inches. Diameter, 3⁄16 inch.
List Price, $36.00 per gross
Packed 1 dozen in a box.   2 gross in a case.
Gross weight, 125 lbs.   Net weight, 90 lbs.   Volume, 4 cubic feet.

Ice Pick
No. 8

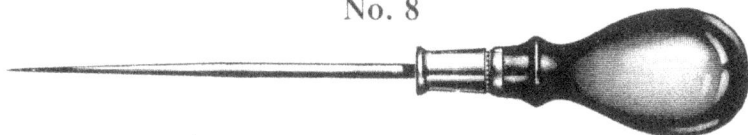

Full tempered steel blade. Nickel plated ferrule.
Red stained varnished handle.
Length of blade, 6 inches.   Diameter, 3⁄16 inch.   2 gross in a case.
List Price, $24.00 per gross
Gross weight, 75 lbs.   Net weight, 40 lbs.   Volume, 4 cubic feet.

Ice Pick
No. 12

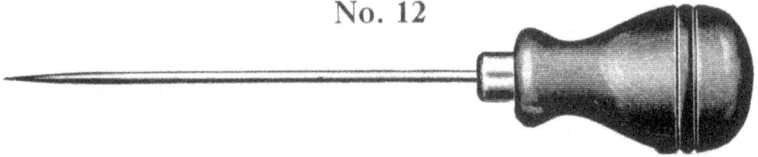

Made of high grade steel—spring tempered. Needle point. Bright finished blade. Nickel plated blade and ferrule. Varnished handle—assorted—Natural—Black —Cherry Red. Length of blade, 6 inches. Diameter, $\frac{9}{32}$ inch.

Packed 1 dozen in a box.   2 gross in a case.   Assorted colors.

List Price, $21.00 per gross

Gross weight, 75 lbs.   Net weight, 40 lbs.   Volume, 4 cubic feet.

Oh-Kay Ice Pick Assortment
No. 360
Two Dozen

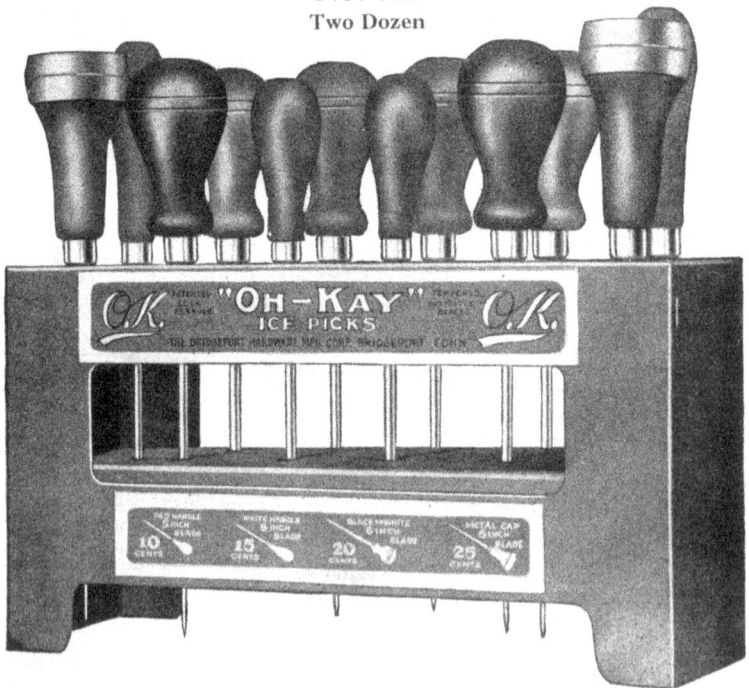

Assortment comprises a variety of shapes and finishes to suit everyone. Attractive display stand of strong construction which makes sales rapidly. Consists of two dozen picks as listed below:

Four No. 3—15 cents each.
Four No. 4—10 cents each.

Twelve No. 12—20 cents each.
Four No. 7—25 cents each.

List Price, $4.40

Weight, 4½ lbs. each.   24 Assortments in case.
Gross weight, 160 lbs.   Net weight, 90 lbs.   Volume, 9½ cubic feet.

The Service Pocket Tool Kit
No. 290

**A Very Useful Outfit for Motorists, Sportsmen, Farmers, Woodsmen, Etc.
Weighs only 5½ Ounces**

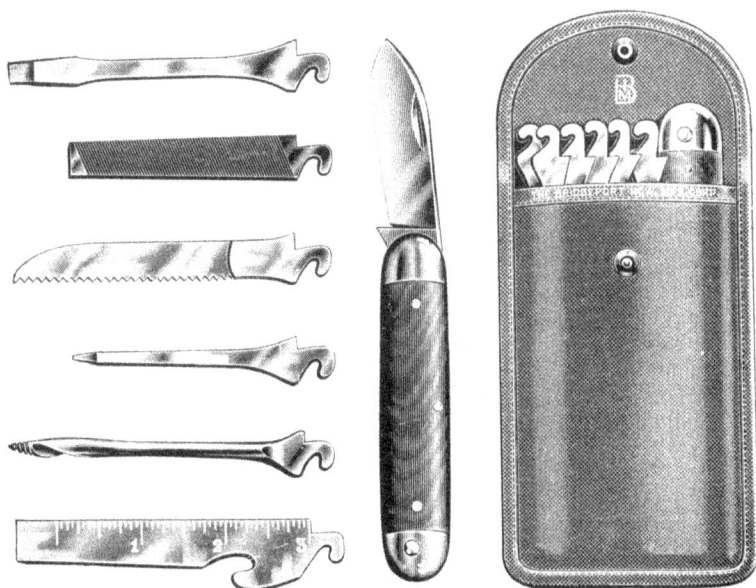

The Service Pocket Tool Kit comprises a good heavy Jack Knife with coco handle, 6½ inches long when open, and a **Chisel, Rule, Bottle Opener, Saw, Screw Driver, Gimlet, Reamer and File.** The knife when closed forms a master handle for the other tools which can be quickly and easily inserted in the end. Tools are full polished. All are packed in a very heavy canvas pocket, khaki color, measuring when closed 3 x 4½ inches.

List Price, $30.00 per dozen

Packed one in a box.   24 dozen in a case.

So Handy Kit
No. 270

Same as **No. 290** shown above but without the saw. Tools full nickel plated, black leather pouch.

List Price, $24.00 per dozen

Packed one in a box.   24 dozen in a case.

Featherweight Tool Chest

PRACTICAL   EFFICIENT   COMPACT

The Set comprises a leatheroid covered Tool Box with tray containing nine high grade serviceable tools and **"Lockgrip"** Master handle.

Saw, Large Screw Driver, Brad Awl, File, Scratch Awl, Gimlet, Reamer, Chisel, Small Screw Driver

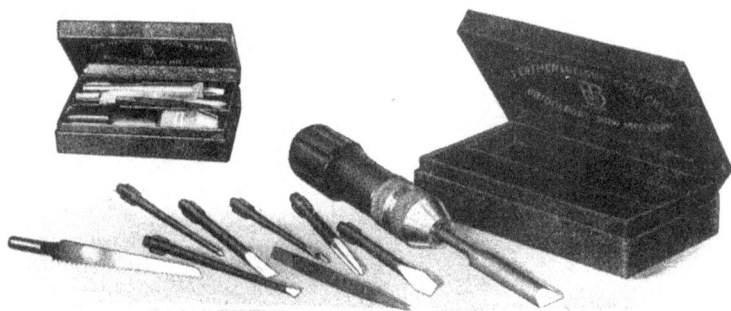

Size, $5\frac{1}{4} \times 3\frac{1}{4} \times 1\frac{1}{2}$ inches.   Weight, 14 ounces.   Length of Tools, 4 inches.

Tools made of fine steel, drop forged, hardened and tempered.

Handle especially designed for close-up work with correct balance and comfortable grip, finished in black rubberoid.

New Steel Chuck is a marvel in simplicity and strength with quick, positive action. A few turns of the knurled sleeve lock any tool rigidly in place. Nickel plated and buffed.

Every set warranted

No. 3   Cardboard case—imitation leather covered—blued finish tools.

List Price, $30.00 per dozen

Gross weight of case, 124 lbs.   Net weight, 98 lbs.

No. 31.   Wooden case—covered with dark red imitation leather—gold lettering. Highly polished tools—Chuck sleeve highly polished, nickel plated and buffed.

List Price, $36.00 per dozen

Gross weight of case, 138 lbs.   Net weight, 110 lbs.
Packed 1 in a box.   10 dozen in a case.
Size of case, 3.2 cubic feet.

Hollow Handle Tool Set
No. 200

Genuine Coco Bolo handle, containing the following four inch tempered steel tools:
**Chisel, Reamer, Small Screw Driver, File, Brad Awl, Saw,
Large Screw Driver, Scratch Awl, Gimlet, Gouge**

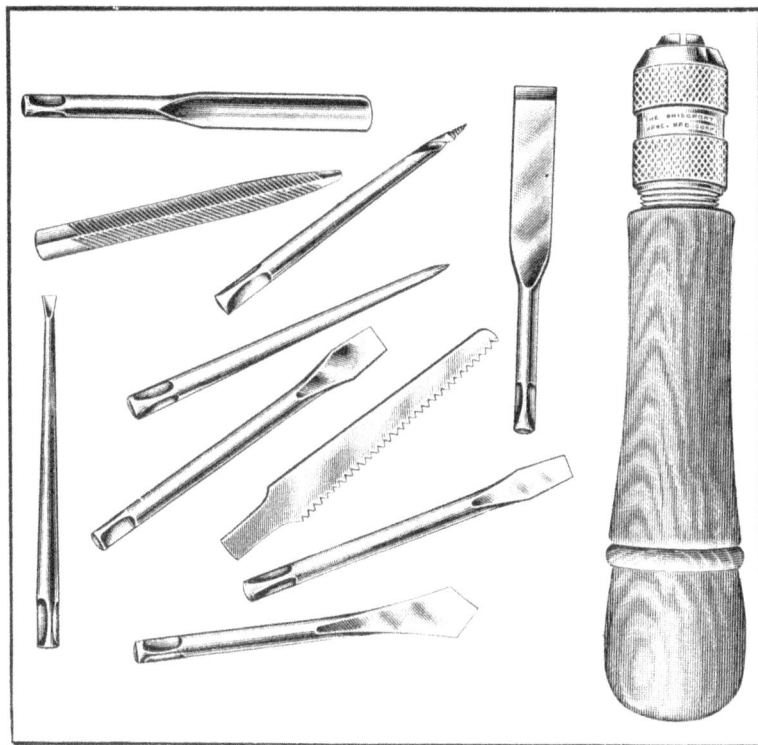

Hand polished Coco Bolo handle.   Chuck of the latest, most approved type, with hardened steel jaws operated by a spring and a heavy knurled steel sleeve. This chuck is fastened securely to the handle by a heavy steel pin.

The very best materials are used in every part, and the construction is very strong and substantial.

No. 200.   Tool Set, ten tools.   **List Price, $30.00 per dozen**
No. 210.   Tool Set, eleven tools.   **List Price,   32.00 per dozen**
No. 220.   Same as No. 200 but with imitation Coco Bolo handle.
                                                   List Price,   24.00 per dozen

No. 210 Tool Set is the same as No. 200, except that an extra 7 inch Keyhole Saw Blade is packed with it.

Length over all, 7½ in.   Weight, 16 oz. each.   Packed one in a box.
6 dozen in case.   Size of case, 1¾ cubic feet.
Gross weight, 87 lbs.   Net weight, 72 lbs.

Utility Household Tool Set

Comprises nine high grade, serviceable tools and a hollow handle which holds all.

File, Saw, Gimlet, Scratch Awl, Chisel, Small Screw Driver, Large Screw Driver, Reamer, Brad Awl

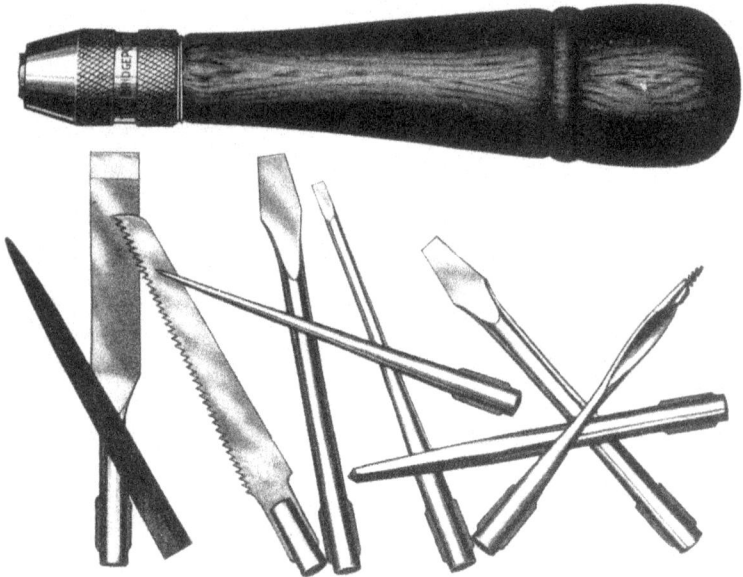

Length over all, 7½ inches.   Weight, 12 ounces.

Tools. Four inches long, forged from fine quality steel, hardened and tempered to give real service.

"Lockgrip" Chuck. Securely fastened to the handle by a heavy steel pin, made entirely of steel. The chuck is very simple and gives quick, positive action. Any tool may be quickly inserted and rigidly held by a few turns of the knurled sleeve.

Handle. Made of hardwood, correctly designed with ample strength, and handsomely finished.

No. 171. Hardwood handle. Mahogany finish. Blued tools.
List Price, $24.00 per dozen

No. 181. Hardwood handle. Mahogany finish. Full polished tools.
List Price, $28.00 per dozen

No. 191. Coco Bolo handle. Highly polished. Full polished tools.
List Price, $30.00 per dozen

Packed ½ dozen in a box.   6 dozen in a case.

The Bridgeport Tool Set

A fine little outfit comprising ten small tools in a hollow handle, as shown below.

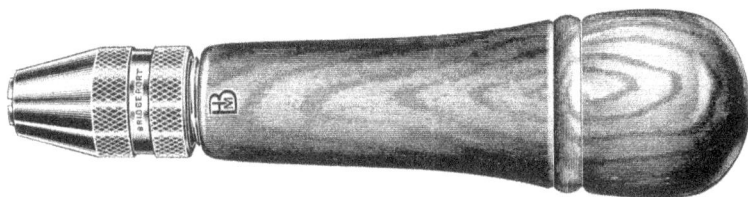

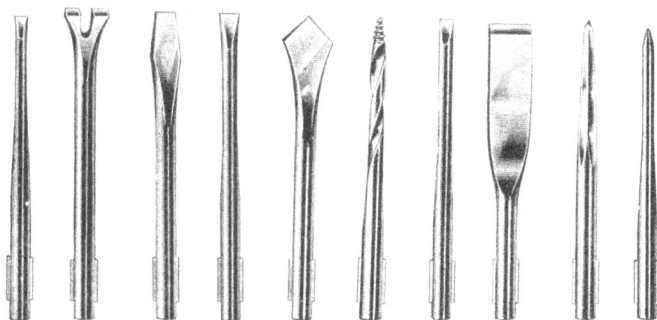

The tools in this set are 2½ inches in length and are of the very best grade.

Handle is made of hardwood stained and polished or genuine coco bolo as listed below, fitted with an exceptionally good chuck. It comprises only two pieces and this simple strong construction provides a chuck which holds each tool rigidly and can be adjusted instantly.

Chuck sleeve is nickel plated and tools are full polished.

No. 282.   Hardwood handle, polished mahogany finish.   Full polished tools.
List Price, $18.00 per dozen

No. 292.   Genuine Coco Bolo handle, beautifully polished.   Full polished tools.
List Price, $24.00 per dozen

Length, 6¼ inches.   Weight, 6½ ounces.   Packed ½ dozen in box.

Favorite Coping Saw

No. 9

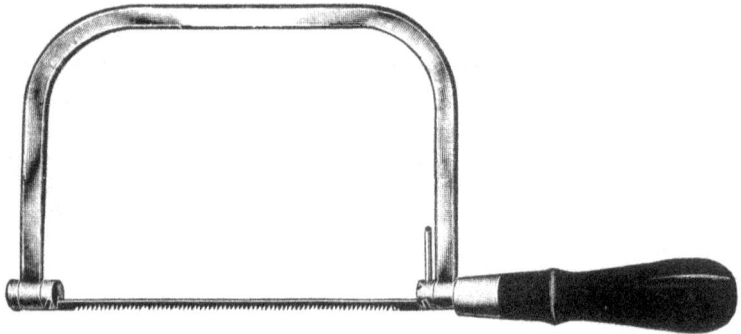

Frame. Heavy steel frame, full polished, nickel plated and buffed.

Handle of hardwood, finished in black rubberoid finish.

Ferrule. Heavy steel nickel plated ferrule threaded on inside.

Blade. Extra fine, $6\frac{1}{2}$ inches in length from pin to pin. Superior in quality and temper. Will cut wood or metal.

A well made, practical, finely finished article throughout.

Complete with two blades.   **List Price, $9.00 per dozen**

No. 19. Nickel plated frame, red varnished handle.   **List Price, $8.00 per dozen**
No. 29. Full polished frame, red varnished handle.   **List Price,   7.20 per dozen**

$\frac{1}{2}$ dozen in box.   12 dozen in case.

Size of case, 4 cubic feet.

Gross weight, 82 lbs.   Net weight, 60 lbs.

No. 200. Blades for **Nos. 9, 19** and **29** Saws.

Length $6\frac{1}{2}$ inches.   **List Price, $4.20 per gross**

All Steel Coping Saw

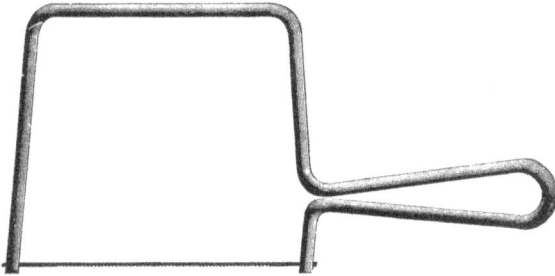

Nickel plated frame made of steel with sufficient spring to keep the blades stretched tightly avoiding breakage.

No. 100.   Packed ½ dozen in a box with one blade in each frame and a dozen extra blades with each.

2 gross in case.   **List Price, $36.00 per gross**

No. 101.   Packed 1 dozen in box—one blade only with each frame.

List Price, $18.00 per gross

Size of case, 2½ cubic feet.

Gross weight, 125 lbs.   Net weight, 100 lbs.

No. 100.   Blades for No. 100, 101 or 210 saws.   **List Price, $1.66 per gross**

1 gross in a box.

Coping Saw
No. 210

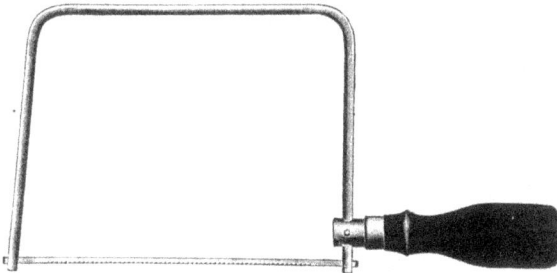

Heavy steel frame ¼ inch diameter. Full nickel plated.

Black wood handle fastened securely to frame.

Cuts either horizontal or perpendicular. 1 dozen extra 6 inch blades with each frame.

Full size, 6 inch.   **List Price, $42.00 per gross**

½ dozen in a box.   2 gross in a case.

Size of case, 4 cubic feet.

Gross weight, 187 lbs.   Net weight, 162 lbs.

Key-Hole Saw
No. 1489

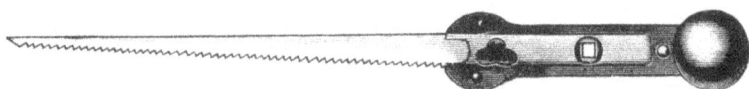

Steel handle, combining a wrench for $\frac{1}{4}$ and $\frac{3}{8}$ inch nuts. Cast steel blade, 7 inch. Thumb screw lock, holds blade securely in handle. The ball shaped end of the handle is a new feature and fits the hand perfectly.

List Price, $4.00 per dozen

1 dozen in a box.   24 dozen in a case.

Size of case, $1\frac{1}{4}$ cubic feet.

Gross weight, 63 lbs.   Net weight, 48 lbs.

Horseshoe Hatchet Gauges

This Attractive Display Card Sells the Goods for the Dealer

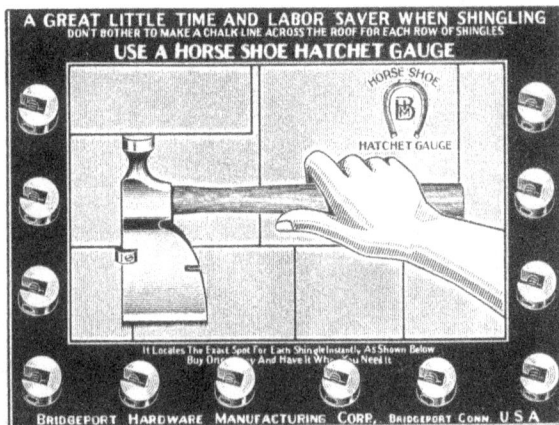

Accurately Gauges Distance between Shingles

No. 20.   Flush head screw, nickel plated finish, 1 dozen in box.
No. 21.   Flush head screw, nickel plated finish, 1 dozen on card.
No. 22.   Square head screw, nickel plated finish, 1 dozen in box.
No. 23.   Square head screw, nickel plated finish, 1 dozen on card.

Packed 1 dozen on a card.   5 gross in a case.   Each card in a container box.

Ticket Punch
No. 30

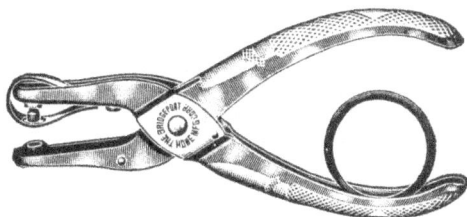

Made of sheet steel with tempered steel stripper and main spring. Scored handles, nickel plated finish. Punches six different shapes as shown below. Packed 6 assorted styles in a box or all round as desired.

Size, 5 inch.   **List Price, $5.50 per dozen**

Stock Dies

For Nos. 30 and 32 Punches

Ticket Punch
No. 32

Same as described above, but with reservoir which holds clippings. This prevents littering of floors.

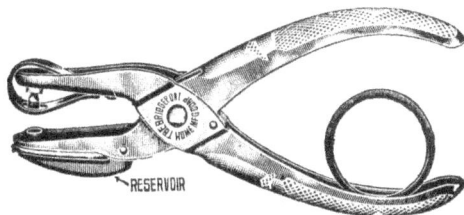

Size, 5 inch.   **List Price, $6.00 per dozen**

Nos. 30 and 32, packed ½ dozen in box.   2 gross in case.

Size of case, 1½ cubic feet.

Gross weight, 60 lbs.   Net weight, 48 lbs.

Bridgeport Office Record Punch
No. 910

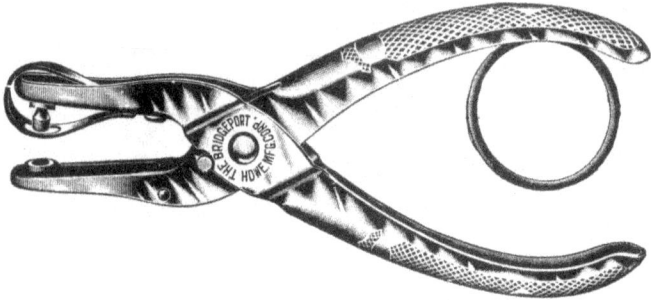

For punching sheets for loose leaf books. Made of sheet steel, full nickel plated and highly buffed. Punches round holes of three sizes, as follows:

Diameter $\frac{1}{8}$ inch $\frac{3}{16}$ inch $\frac{1}{4}$ inch

This punch is designed for light work only. We do not recommend them for use on thick material and they are not warranted.

List Price, $7.20 per dozen

Specify size of die when ordering.

Packed $\frac{1}{2}$ dozen in box.   2 gross in case.

Toro Ticket Punch
No. 42

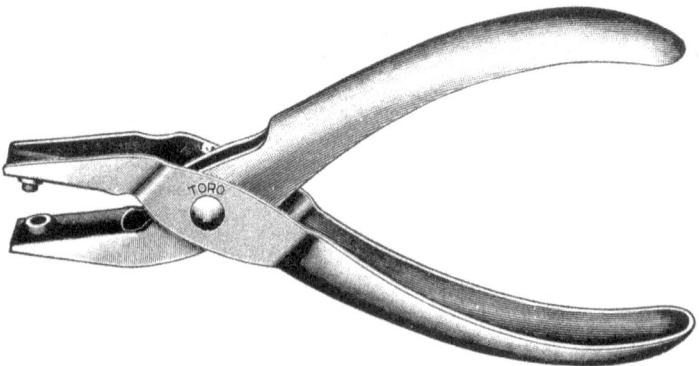

A substantial, well finished steel punch that can be retailed for fifteen cents. All round dies—bright polished finish—five inches long.

List Price, $24.00 per gross

$\frac{1}{2}$ dozen in box.   2 gross in case.

Size of case, $1\frac{1}{4}$ cubic feet.

Gross weight, 56 lbs.   Net weight, 42 lbs.

Economy Book Support
No. 40

Size, 4¾ x 5½ inches.

Made of Extra Hard Roltem Steel

These Book Supports solve the problem of holding a row of books upright on a desk top, shelf or table in the home or office. They are used in pairs as shown below. Very useful in schools and libraries for partially filled shelves.

The original Economy Book Support which has been widely known and used by the trade for many years. It should not be confused with cheaper imitations which are of uncertain quality. Look for our brand and trade mark which appear on every genuine Economy Book Support.

No. 40.   Roltem sheet steel, nicely japanned.   **List Price, $12.00 per hundred**

Packed only in wooden boxes containing 100 holders.

500 in case.

Size of case, 1¾ cubic feet.

Gross weight, 200 lbs.   Net weight, 190 lbs.

Pointing Trowels

No. 160. 5½ inch bright steel blade. Forged steel shank. Natural wood handle. Packed 2 dozen in box. Brass ferrule. **List Price, $15.00 per gross**

No. 16. 5½ inch. Tempered steel blade nicely polished. Dark red polished handle; polished steel shank, nickel plated ferrule. Packed ½ dozen in a box.

List Price, $24.00 per gross

Nos. 16 and 160, packed 2 gross in case.

Size of case, 2½ cubic feet.   Gross weight, 97 lbs.
Net weight, 80 lbs.

Brick Trowels

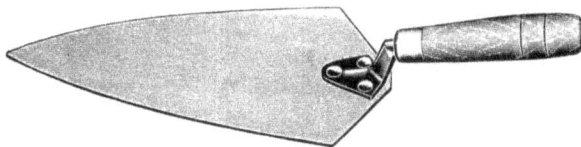

No. 170. 10 inch bright steel blade. Steel shank. Natural wood handle. Steel ferrule.

List Price, $24.00 per gross

No. 17. 10 inch tempered steel blade, nicely polished. Dark red polished handle, polished steel shank, nickel plated ferrule. **List Price, $48.00 per gross**

Nos. 17 and 170 packed ½ dozen in box, 2 gross in case.

Size of case, 4½ cubic feet.   Gross weight, 168 lbs.
Net weight, 144 lbs.

Plastering Trowels

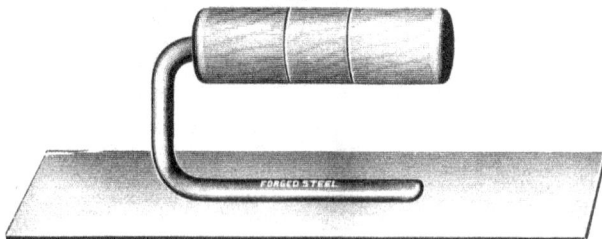

No. 180. 10 inch bright steel blade. Forged steel shank. Natural wood handle.

List Price, $36.00 per gross

No. 18. 10 inch tempered steel blade, nicely polished. Dark red polished handle, forged steel, polished shank. **List Price, $48.00 per gross.**

Nos. 18 and 180 packed ½ dozen in a box, 1 gross in case.

Size of case, 4½ cubic feet.   Gross weight, 148 lbs.
Net weight, 120 lbs.

Bridgeport Putty Knife
No. 135

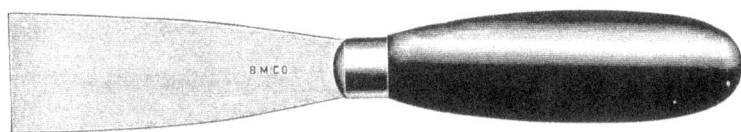

A very good low priced knife

Tempered cutlery steel blade 3½ inches long, red stained handle, brass cap ferrule. Width of blade, 1⅛ inches.

1 dozen in a box.   2 gross in a case.

List Price, $15.00 per gross

Scratch Awl
No. 200

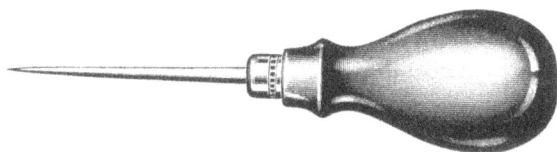

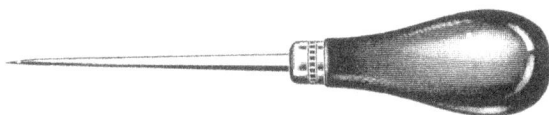

Length of blade, 4 inches.   Diameter, $\frac{3}{16}$ inch.   Needle point.

1 dozen in a box.   24 dozen in a case.

List Price, $1.80 per dozen

Scratch Awl
No. 400

Length of blade, 4 inches.   Diameter, $\frac{3}{16}$ inch.   Needle point.

1 dozen in a box.   24 dozen in a case.

Gross weight, 44 lbs.   Size of case, 1¾ cubic feet.   Net weight, 21 lbs.

List Price, $1.20 per dozen

Scraping Knife
No. 129

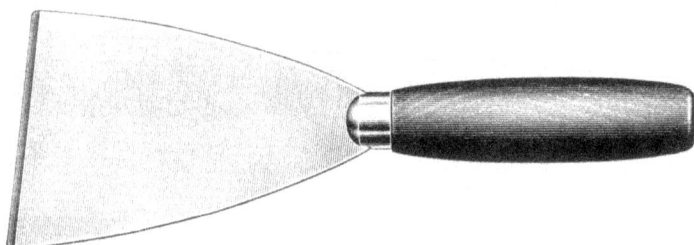

Bright Steel Blades.   Bevel Edge.   Red Varnished Handle.

Nickel Plated Ferrule.   Width of Blade, $2\frac{3}{4}$ inches.

3 gross in case.   Net weight per gross, 27 lbs.

List Price, $21.00 per gross

The P-B Scraper
No. 125

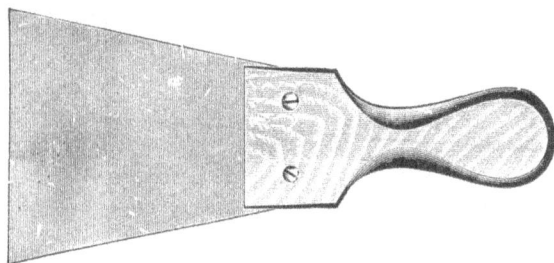

For removing old paint and wall scraping, also used extensively by bakers for cleaning pans, etc., and by candy makers.

The P-B Scraper is fitted with rust-proof screws, an exclusive feature. Unlike the ordinary scraper it can be kept clean and sanitary by washing without rusting out the screws.

The blade is made of roll tempered steel.   Beech handle.   Natural finish.

Made in three widths of blade, 3 inch, $3\frac{1}{2}$ inch, 4 inch.

Two gross in case.

Gross weight, 120 lbs.   Size of case, 4 cubic feet.   Net weight, 100 lbs.

List Price, $36.00 per gross

Tools

A line of popular priced tools designed by practical, experienced radio mechanics. Comprises just the necessary and special tools for use by radio builders, experimenters, etc., and for the adjustment and repair of factory built sets.

All tools packed with attractive display cards or stands.

Radio-Lectric Screw Drivers
No. 45 ASSORTMENT

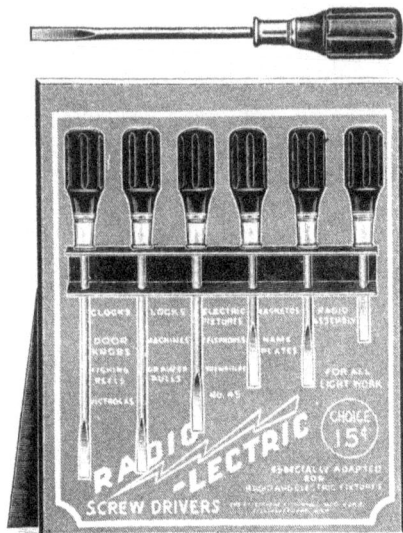

Especially adapted for radio assembly and electric fixtures. Parallel sides of point permit reaching recessed screws.

Assortment comprises,

2—2 inch   2—4 inch
4—3 inch   4—5 inch

Packed, 24 assortments in case.

Size of case, 3¾ cubic feet.

Gross weight, 51 lbs.

List Price, $1.80 each

Net weight, 27 lbs.

No. 465
Radio-Lectric Reamer

An indispensable tool in radio work. For enlarging drilled holes to the proper size. Only one size of drill need be used. Operates perfectly on bakelite, hard rubber, etc. One-half dozen with display.

Packed 24 dozen in a case.

List Price, $3.00 per dozen

Size of case, 2¼ cubic feet.

Gross weight., 60 lbs. Net weight., 30 lbs.

No. 450
Radio-Lectric Wiring Plier

Forms perfect evelets of any size quickly and easily. Good wire cutter on both sides of the tool giving a clean cut.

One half dozen with attractive display card.

Packed, 24 dozen in a case.

List Price, $4.20 per dozen

Size of case, 3¼ cubic feet.

Gross weight., 84 lbs. Net weight., 51 lbs.

Radio-Lectric Plier
No. 440

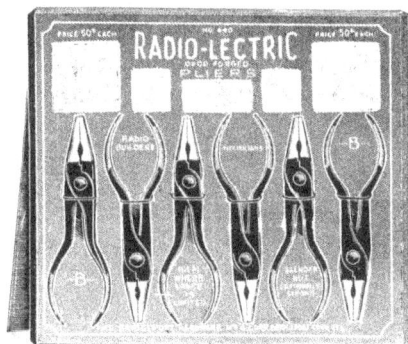

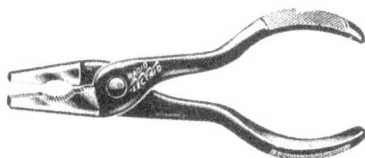

Designed especially for radio and electrical work. Slim nose shaped to reach places where space is limited. Wire cutter between jaws. Radio-Lectric Pliers are drop forged from a quality steel and are fully guaranteed. One-half dozen pliers on attractive display card.

Five drawings showing practical uses.

Packed, 12 dozen in a case.

List Price, $6.00 per dozen

Size of case, 2 cubic feet.

Gross weight, 60 lbs. Net weight, 33 lbs.

No. 445
Radio-Lectric "Crow Bill" Plier

An ideal plier for all radio and electrical work. Long thin semi-round nose for fine jobs where other types of pliers are useless. Just right for radio wiring and enables the user to form perfect eyelets. Drop forged. Warranted.

Attractive display card illustrating practical uses with each one-half dozen. Packed 12 dozen in a case.

List Price, $6.00 per dozen

Size of case, 1 cubic foot.

Gross weight, 50 lbs. Net weight, 25 lbs.

No. 116
Radio-Lectric End Cutting Nipper

A needed tool in radio work. Cuts all kinds of wire. Correctly designed, well made and durable. Drop forged and guaranteed. Packed one-half dozen in a box without display card—24 dozen in a case.

List Price

5½ inch size,   60c each.
6½ inch size,   70c each.

Size of case,
5½"-3¼ cubic feet. 6½"-3¾ cubic feet.
Gross weight, 156 lbs. Net weight, 120 lbs.
176 lbs.   144 lbs.

Radio-Lectric Wrench Set
No. 475

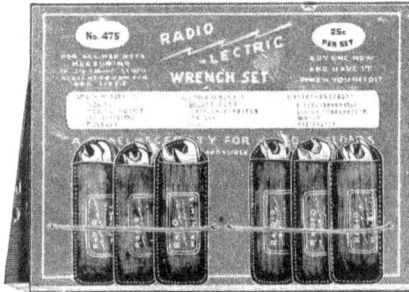

A real necessity for radio builders, repairmen, electricians and auto mechanics. Adjusts nuts in places impossible to reach with other tools. For all hex. nuts measuring $\frac{1}{4}$ in., $\frac{5}{16}$ in., $\frac{3}{8}$ in., $\frac{7}{16}$ in. and $\frac{1}{2}$ in., with alligator jaw for odd sizes. Each set in genuine leather case, twelve sets with a display card.

Packed 36 dozen in a case.

List Price, $3.00 per dozen

Size of case, 2 cubic feet.

Gross weight, 72 lbs.   Net weight, 45 lbs.

¼"   ⁵⁄₁₆"   ³⁄₈"   ⁷⁄₁₆" ½"   ODD SIZES

No. 442
Radio-Lectric Nut Forcep

A perfect grip on all size knurled, hex. and square nuts. Fine for cotter pin work, also for pins and horseshoe fasteners on gas engines. Radio builders will find it the only practical tool for knurled nuts and the Nut Forcep fits all sizes. Radiotricians—Auto Mechanics—Electricians.

Drop forged. Warranted. Six with attractive display. Packed 12 dozen in a case.

List Price, $7.20 per dozen

Size of case, 2 cubic feet.

Gross weight, 55 lbs.   Net weight, 36 lbs.

No. 405
Radio-Lectric Countersink

For shaping holes in hard rubber, bakelite, fibre, wood, etc., to fit the heads of flat head screws. A necessary tool for the radio builder. One dozen with counter display.

Packed 36 dozen in a case.

List Price, $1.80 per dozen

Size of case, 1¾ cubic feet.

Gross weight, 56 lbs.   Net weight, 27 lbs.

Radio-Lectric Socket Wrenches

Sockets snap on to shank and assembly is quick. Hollow stem to take projecting screws. Strong but thin walls of sockets permit reaching nuts where space is limited. Tight connections are absolutely necessary for perfect reception. The Radio-Lectric Wrench saves time and labor. Also needed by electricians, auto mechanics, telephone repairmen and others.

No. 480 Set

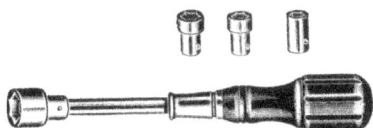

Sizes $\frac{1}{4}$-$\frac{5}{16}$-$\frac{3}{8}$ inch.
List Price, $6.00 per dozen
One set to a box.   12 dozen in case.
Size of case, 3 cubic feet.
Gross weight, 52 lbs.   Net weight, 24 lbs.

No. 485 Set

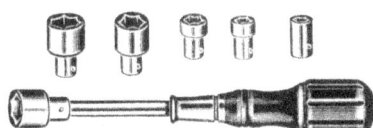

Sizes $\frac{1}{4}$-$\frac{5}{16}$-$\frac{3}{8}$-$\frac{7}{16}$-$\frac{1}{2}$ inch.
List Price, $9.00 per dozen
One set to a box.   12 dozen in case.
Size of case, 3 cubic feet.
Gross weight, 63 lbs.   Net weight, 36 lbs.

No. 410

Radio-Lectric Center Punch

An important tool in the making of radios. Panels must be prick punched as a starter for the drill. One dozen on display card. Fully guaranteed. Packed 36 dozen in a case.

List Price, $1.80 per dozen

Size of case, $1\frac{3}{4}$ cubic feet.

Gross weight, 62 lbs.   Net weight, 36 lbs.

No. 430

Radio-Lectric Scriber

Lays out panels and baseboards accurately. Saves time and insures neatness. The finest of steel, fully tempered, is used in Radio-Lectric Scribers. Needle point on each end for marking hair line and will fit the very smallest screw hole. One dozen on card. Guaranteed.
Packed 24 dozen in a case.

List Price, $3.00 per dozen

Size of case, 1 cubic foot.

Gross weight, 23 lbs.   Net weight, 12 lbs.

Radio-Lectric Combination Wrench Set

No. 490

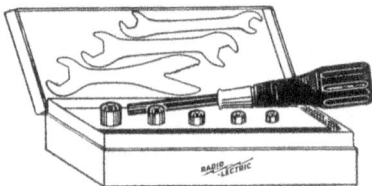

A compact set of wrenches for all hex. nuts from ¼ inch to ½ inch. Neatly packed in partitioned case with dark blue covering. Socket wrenches in five sizes with one handle. Three flat wrenches which adjust nuts impossible to reach with other tools.

Packed 12 dozen in a case.

List Price, $12.00 per dozen

Size of case, 2½ cubic feet.

Gross weight, 100 lbs.   Net weight, 50 lbs.

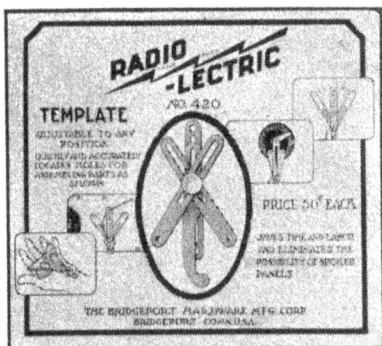

Radio-Lectric Template
(Adjustable)
No. 420

Indicates the exact location for the drilling of holes in panel or baseboard of any radio set. Fits perfectly on any part and is then locked in position by the thumb nut. Transferred to the panel where the part is to be located, mistakes in drilling are avoided. A great time and trouble saver. Six templates with display card. Packed 24 dozen in a case.

List Price, $6.00 per dozen

Size of case, 2½ cubic feet.

Gross weight, 61 lbs.   Net weight, 24 lbs.

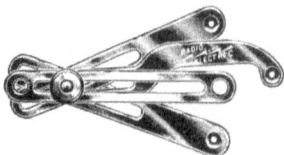

Radio❧Tool Kit
No. 495

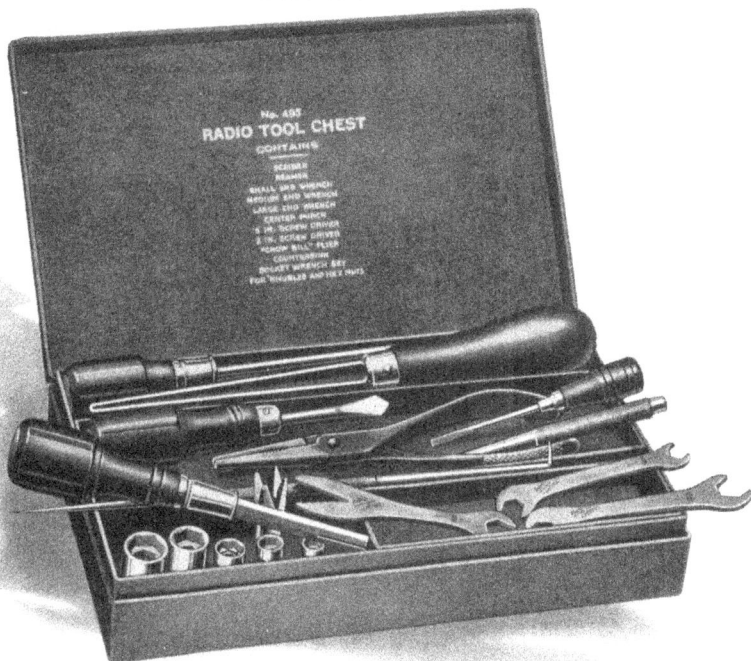

Contains ten practical tools designed by men who knew exactly what was required for radio work.

All in a well-made, very attractive dark blue and gold cardboard case as illustrated.

Retails at $3.00

Contents:

Three inch screw driver for small work
Five inch screw driver
Scriber for layout work
Reamer for enlarging holes (eliminates the necessity of a full set of drills)
Countersink
Plier—long thin semi-round nose
Center punch
Five size socket wrench set for hex. nuts
(One handle—sockets snap on to shank)
One socket for all size knurled nuts
Small end wrench for $\frac{1}{4}$ inch and $\frac{5}{16}$ inch hex. nuts
Medium end wrench for $\frac{3}{8}$ inch and $\frac{7}{16}$ inch hex. nuts
Large end wrench for $\frac{1}{2}$ inch hex. nuts and alligator jaw for all odd sizes

Packed 6 dozen in a case.   Each kit in a container box.
Gross weight, 136 lbs.   Net weight, 90 lbs.   6½ cubic feet.

List Price, $36.00 per dozen

Radio-Lectric Socket Wrench Set
For Hex and Knurled Nuts
No. 481

Contains three sockets for hex. nuts, $\frac{1}{4}$, $\frac{5}{16}$, and $\frac{3}{8}$ inches and three sockets for all styles and sizes of knurled nuts. All sockets fit on the one handle and shank. This set of socket wrenches is indispensable for the radio builder.

One set in a box.   Packed 12 dozen in a case.

List Price, $12.00 per dozen

Gross weight, 50 lbs.   Net weight, 35 lbs.

Volume, $1\frac{1}{4}$ cubic feet.

No. 483
Radio-Lectric Socket Wrench Set For Knurled Nuts

No. 491
Radio-Lectric Combination Wrench Set

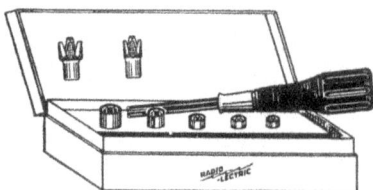

Three sockets which fit all styles and sizes of knurled nuts used on radio sets. Sockets snap on to shank and assembly is quick. Design of sockets insures a quick, positive grip without marring. Sockets will not slip or clog.

One set in box.   12 dozen in case.

List Price, $9.00 per dozen

Gross weight, 45 lbs.   Net weight, 27 lbs.

Volume, $1\frac{1}{4}$ cubic feet.

A complete set of wrenches for all hex. and knurled nuts. Five sockets for hex. nuts from $\frac{1}{4}$ to $\frac{1}{2}$ inch by sixteenths, and three sockets for knurled nuts. Neatly packed in partitioned case with dark blue covering.

List Price, $15.00 per dozen

12 dozen in case.   Volume, $2\frac{1}{2}$ cubic feet.

Gross weight, 90 lbs.   Net weight, 40 lbs.

Index

Notice—Table showing Net Weight and Cubic Measurement of All Screw Drivers on Pages 42–45 inclusive.